KNIGHTS OF THE
CORPORATE ROUNDTABLE

The Success Story of TCDI

Mary Walsh

Mary Walsh

REGENT PRESS
Berkeley, California

Copyright © 2023
Published by Regent Press
Berkeley, California
Library of Congress Control Number: 2023938113

Printed in the United States of America
By DavCo Advertising, Inc., Kinzers, Pennsylvania

ISBN 10: 1-58790-658-9
ISBN 13: 978-1-58790-658-9

Cover by Tom Gonzalez
tomprints.com

Also by Mary Walsh

The Curse of Jean Lafitte
American Posse
Memories of 9/11
Plenty of Fish in the Ocean State
Once Upon a Time in Chicago
His Second Chance
Wounded but not Dead
Where or When
Fine Spirits Served Here
You Deserve Better
Life Lessons for My Kids
Stable of Studs
Dragon Slayer
Catch a Break

CONTENTS

FOREWORD

By Bill Johnson, CEO

Somewhere in the dark reaches of my mind, in thinking about what 35 years of building TCDI means on a personal and professional level, the idea of putting together a book floated into consciousness. Not sure what the original thought covered, but in trying to clarify my thoughts, and with talking with Mary, we came upon the idea of writing to celebrate our success. This book is not, however, simply a chronology of the history of TCDI. Rather it's a collection of vignettes--vignettes about the people who have built this company, vignettes about the lessons learned through various projects and clients (some of which were hard but those teach the most), and vignettes about how TCDI approaches its work.

The key to understanding TCDI, and what I hope you will gather from this book, is that it is all about the people. When we established TCDI in 1988, it wasn't because we had a great software idea or an original approach to solving a business problem. It wasn't because we thought we were better than everyone else, but it was only a matter of time before others recognized that and would beat down our doors with work. It was because Jerry Eatherly and I, the founders, had the most satisfaction and fun exchanging ideas and thoughts with what can best be described as exceptionally smart people. On one fateful night at our local watering hole, we decided that the way to ensure that ability to find and work with smart and challenging people was to start our own company.

The idea of TCDI being a people-centric company may sound strange to those who know me personally… after all, I'm your classic introvert. There isn't a party or social gathering I haven't tried to figure out how to avoid. Definitely not the life of the party. But at the same time, I tended to collect people into the inner circle that were diverse, from all spectrums of life. Starting with Jerry…

Jerry is retired Army who got into computers after the Vietnam War as a combat helicopter pilot who needed to move to a new career. We met at a consulting firm programming business computer graphics for decision-makers. As you can imagine Jerry was a bigger-than-life chopper cowboy with endless stories but one with incredible brains to back up that bravado. In a lot of ways, his personality was everything mine wasn't…a perfect combination of yin and yang.

Our initial team was from all over the map. A young lieutenant in the army looking for his first civilian job following the first Iraq war where he wrote logistical databases in his spare time. A grizzled veteran of working with the intelligence agencies on information security, way before the days when information security wasn't cool but deadly serious. An ex-Army Non-Commissioned Officer who was Hollywood's vision of what an NCO was: the one guy who could make any organization work, but who was also a wiz at information management. A guy straight from college who was so analytical it took him a year to decide on his first car but could out-code anyone. An Air Force Academy grad who flew an F-111 over Vietnam (of course, if you ask Jerry, he wasn't a part of the "real" war happening down low where the choppers flew).

Diverse was an understatement. But it instilled in all of us the ability to exchange ideas, to listen, to share responsibility and accountability, and to truly lead instead of simply manage. We were accountable to each other and to our clients, not to outside venture capitalists or investors. Our personality was formed early and developed into our core values.

The people that make up TCDI have changed over the years, and grown significantly, but our personalities and convictions have always remained.

This book is about how our personality has taken shape and developed, how it has translated into our success, and our management philosophy. If you take the concepts of smart thinking, shared responsibility, mutual respect, equality, creative thinking, and commitment to clients, remove overreaching egos and internal political thinking, and hire according to those principles, you foster an environment that attracts the smartest and brightest. Then you build management philosophies, such as Lean Six Sigma, to guide the efforts and focus of the company. It doesn't matter what industry you practice in; you have the blueprints for success. That is what this book is about in practical terms…how TCDI was built to become a blend of the best dedicated to a central idea of what it means to be successful for all the right reasons.

Or, to blatantly steal from the legend of King Arthur, how the people of TCDI became the Knights of the Corporate Roundtable.

1

Who is **TCDI?**

Fortune 100 Companies. American Lawyer 100 law firms. Government agencies. The largest and most complex litigations in U.S. history.

What do all of these have in common?

TCDI has been entrusted to serve as the partner and service provider for some of the largest companies, law firms, and litigation dockets in the United States.

Headquartered in Greensboro, North Carolina, with other offices in Cleveland, Ohio, and Purchase, New York, TCDI's client base is diverse. As client needs have evolved, so has TCDI's expertise in developing the highest quality software, security, e-discovery services, and network architectures. Our products and services are superior because we seek and retain only the best and brightest minds in the business. That is the core of TCDI's culture and reputation. Our clients deserve nothing less. With 175 employees, TCDI has an excellent reputation with stellar client references.

TCDI maintains one of the longest durations in business as a single eDiscovery entity. (eDiscovery is the electronic aspect of identifying, collecting, and producing electronically stored information in response to a request for production in a lawsuit or investigation.) The average client tenure at TCDI is 10-15 years. International companies trust TCDI and want to stay with them. Over 500 international law firms and corporations use TCDI technology and services every day.

For over 35 years, our knowledgeable team of experts in legal technology and information security has delivered uncompromising service and best-in-class technology. Our people-centric method makes us unwaveringly accountable, responsive, and reliable. And easy to work with.

TCDI serves such industries as pharmaceutical, automotive, consumer goods, food and beverage, health insurance, tobacco, government, and energy. All of these businesses' communication is discoverable in a litigation case. This includes texting, social media chats, and workplace collaboration.

TCDI started small but with big ideas. We built products and teams around the highest standards of data security and privacy. Simply put, TCDI puts passion for data to work. Our engineers and developers used their deep understanding of our clients' needs and challenges. Through this, we created some of the best eDiscovery platforms and processes in the market. TCDI excels at forensics, processing, hosting, and review, and works with many large and small corporations, law firms, and government agencies.

Our applications adeptly handle the unique demands of large-scale mass tort, common complaint, and other complex, multi-case litigation.

If a company has a problem, we provide a solution.

According to Japanese legend, the Crane is a symbol of luck. As the original corporate logo for TCDI, the origami crane represents good luck for TCDI and our clients. It also reflects the Japanese heritage of one TCDI co-founder (Bill) and the aviation background of the other (Jerry). Origami is the art and science of folding paper into a unique design. This ancient Japanese art of transforming a flat sheet of paper into any shape requires great technical skill, discipline, and logical thinking. We at TCDI apply that same process to transform significant data into valuable knowledge management systems for our clients.

2

FROM TERRAPINS TO TERABYTES

As an undergrad at the University of Maryland in the late 1970s, Bill Johnson majored in architecture with aspirations to erect the next Willis Tower. He quickly learned that he was more interested in computer programming than designing and constructing buildings. Coding danced through his dreams. His career path was in sight.

Upon graduation in 1981, Bill landed his first job with Decision Resources Corporation (DRC), a government consulting and executive information systems company. As a programmer, he did basic coding, led a team of developers, maintained client interaction, and managed programs. By his mid-twenties, Bill was making twice as much money as his fellow Terrapins. He enjoyed being presented with a technical problem and providing a solution. He worked at DRC for three years, then had a stint at EDS for a year developing electronic photo-type setting management systems under the Department of Energy. Bill realized that he didn't like the big company culture of EDS and went back to DRC.

While working at DRC for the second time, Bill met Jerry Eatherly. Jerry was a retired Lieutenant Colonel in the U.S. Army who flew AH-1G attack helicopters over Vietnam. To keep his mind engaged, Jerry joined the ranks of computer programmers at DRC.

In 1988, after a long week at DRC, Bill and Jerry shared some beers one night. After discussing the mediocre state of affairs in the tech industry, they decided they could do better. As a company and as managers. Technology Concepts & Design, Inc. (TCDI) was born in Manassas, Virginia.

With a staff of three other people, Bill and Jerry landed TCDI's first contract with the George H W Bush Inaugural Committee. The following year came the Development of the Arms Control Negotiation Research Database for Arms Control and Disarmament Agency. The highly classified project involved messaging traffic among negotiators in the arms treaty. TCDI built systems for analysts to search terabytes of tech data within secure messages.

Two years later, Bill secured a meeting with the prestigious New York law firm Debevoise & Plimpton LLP. The billion-dollar international law firm was founded in 1931 by Harvard Law alumnus Eli Whitney Debevoise and William Stevenson who was the American Ambassador to the Philippines under President Kennedy.

Bill had been working with Espe, a staff attorney, at Debevoise & Plimpton. She had set up a meeting with one of the partners at the law firm, which meant it was a vital meeting for the future success of TCDI. Dressed in his finest suit and an overcoat (luckily, since it was winter), Bill ran for a plane at Dulles airport. Dashing up the escalator, he heard that wonderful ripping sound of his pants splitting. Thank goodness for the overcoat! Not thinking through what might be the best course of action, but knowing canceling the meeting was out of the question since it had taken months to set up, not to mention explaining why he had to cancel seemed unmentionable, Bill forged ahead. The warm plane didn't help, which made the overcoat close to intolerable, but modesty dictated he keep it on. Arriving at the law offices in the heart of Manhattan, Bill confided his predicament to Espe. She gave Bill the use of her office to hide out in and took his trousers to an alteration shop around the corner (New York was the perfect place to be, anything could be found on the block). Locked in her office in an overcoat with no trousers may be a look one could find in Central Park, but not welcome in a prestigious office environment. Thus, Bill waited patiently for the return of his trousers. He never envisioned his first big-time meeting in New York like this. He couldn't thank Espe enough for the rescue.

Bill's short-term embarrassment was superseded by long-term success. Debevoise & Plimpton signed a contract as TCDI's first litigation client. TCDI developed one of the first online document/image retrieval systems for them. Three decades later, Debevoise & Plimpton is still a client of TCDI.

Unfortunate things happen. It's a fact of life. Bill had a problem and he solved it.

In 1998, Bill opened an additional office in Winston-Salem, North Carolina, to work on a project for R. J. Reynolds. In 2000, the second office was relocated to Greensboro, North Carolina, as more clients needed large-scale database management.

As the CEO of TCDI, Bill provides vision, leadership, and multiple decades of industry knowledge. For over three decades, he has thoughtfully guided TCDI's evolution and expansion to meet the needs of clients. His commitment to building long-term partnerships and developing innovative solutions has led to tremendous growth and expanded service offerings. He clearly communicates his vision, empowers his team to execute, works alongside them to address issues that arise, and ensures they have the resources to be successful.

TCDI offices and staff locations

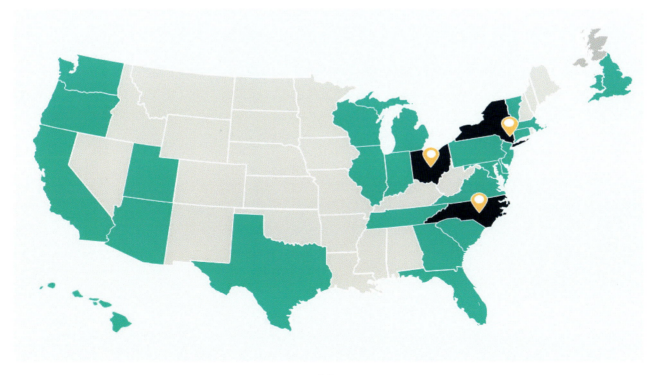

3

KNIGHTS OF THE CORPORATE ROUNDTABLE

In the 12th Century, legendary King Arthur forged a Round Table for his knights in pursuit of the Holy Grail. Each knight, who ranged from sovereign royals to minor nobles, had a fair and equal stake in the quest.

At TCDI, all of us are *knights*. From various backgrounds, everyone has equal say and equal opportunity to voice their solution to a problem. We all metaphorically sit together to symbolize oneness, equality, and unity for the best interest of TCDI. With this type of completeness, we have the confidence to reach new heights. The innate hesitation of approaching upper management that subordinates have in most companies is gone. At TCDI, everyone is encouraged to present opportunities and analyze problems. It's safe to speak up and okay to make mistakes.

King Arthur also established the *Pentecostal Oath* where his knights were expected to uphold his expectations and be rewarded for their chivalrous deeds. Bill follows a modern-day version of this Code of Conduct where he has high expectations for us and remunerates us greatly for success. Instead of land or a horse, Bill offers luxury corporate cars, trips, and tickets to sold-out events. When a staff member had to work into the night on a project, Bill thanked him and told him to take his wife away for the weekend—on TCDI's dime. Few CEOs would do this.

Most companies have standard organization charts full of multi-colored squares and lines. We do it differently. Our organization chart is a circle, like King Arthur's Round Table. Instead of a pyramid hierarchy peopled with

managers and subordinates, we have homogeneous sections. Everyone has the same empowerment to bring change. We are all leaders.

We are encouraged to state our ideas to solve problems. Some ideas may stick and others may not, but everyone has the chance. The only caveat is that we check our ego at the door. We can walk into Bill's office and state our idea without feeling intimidated. His door is always open for employees to take a seat on his couch and talk to him. Showing care and interest, Bill continually gives us the security of being heard. One of the best sources of information at TCDI is our employees.

Although it took a few years for this even-handed organization chart to evolve at TCDI, eventually all departments came around. As a result, we had to adjust with the new organization chart. Each department had an evolution (and some hesitancy) to get to the end result.

In addition, Bill distributed copies of the book *Turn the Ship Around* by David Marquet to all employees. Captain Marquet successfully transformed the *USS Santa Fe* in less than a year in 2000 from the worst-performing submarine in its fleet to the best. The book presents a different approach to leadership, using a "leader-leader" model instead of a "leader-follower" model. We quickly latched onto this concept. The executive team is noticeably absent from our round chart. We feel empowered to say, "I'm going to do…" instead of "Can I…?" As a whole, we take care of everything as Bill steps back and lets us figure out solutions.

Even though the new organization chart has been in effect for several years, TCDI managers still jump in as necessary to work with staff to complete a project. The manager's job is not only to issue orders but he or she is expected to remove roadblocks for us. Managers serve their departments, not the other way around. As a result, everyone has evolved to be better and that mentality paid off.

Often, we have said, "How can we make this better for a client?" and worked together to find the best solution. In the tech industry, it's tough to find a company where staff always rely on each other to help a client succeed. We don't think twice about it.

Our talented staff takes ownership and everyone at TCDI triumphs.

4

TRUST IS EXPECTED

At TCDI, Bill Johnson does not believe in micro-managing. He prefers to trust us to do our jobs which are good for our employees, managers, and the company as a whole.

A great leader shows sincere enthusiasm for his business, its products, and its mission. Bill holds himself to the highest level of excellence and expects nothing less from our team. He works alongside us pulling all-nighters and addressing issues that arise. He stays ahead of industry trends and his expertise challenges and, as a result, sharpens us. Whether acknowledging his mistakes, giving credit for accomplishments, or putting quality above the bottom line, Bill exhibits integrity in all that he does.

However, more important than his technical knowledge and enthusiasm is how he treats us. We are a family to Bill. He trusts us and he gives us the latitude we need to thrive and grow. He genuinely listens. He stays involved. Transparency matters to him. He speaks with authority and his words, though softly spoken, command the respect that is given to someone who delivers on what they say and whose ideas are proven true in the business world. His priority is his people and he goes above and beyond to treat us well and take care of us.

Many employees at TCDI have personal stories of Bill reaching out to help them. One long-time employee liked the freedom to work remotely. When he approached Bill about working from a newly-bought beach house, Bill openly supported him. Bill trusts that his staff will do the right thing and that trust is reciprocated.

An innovative culture was created by hiring the best people for the job and fashioning an environment of mutual trust. Our ideas, thoughts, and opinions are fully embraced company-wide.

In 2013, when Jerry Eatherly retired, the opportunity arose for Bill to reward the rest of us. He didn't want only the top 10 key employees to earn benefits but wanted everyone to be involved. This made sense because the value of TCDI lay in the involvement of our employees. If anyone was excluded, the whole team could suffer.

Bill made the ultimate statement of his commitment to us when he made TCDI an ESOP company. An Employee Stock Ownership Plan is a stock bonus plan for staff. Not only did Bill want to profit from the success of the company, but he also wanted us to reap the benefits. Bill recognized there would be good times and bad, as occurs in all businesses. He knew he might have to lead us through some years when profits might flatline. Regardless, we were all in it together.

It's almost unheard of to find an ESOP company in the tech industry. As employees, we are interested in growing the company because we have more vested in it. We want TCDI to succeed because we reap the benefits. No one wants to see the company fail because this is our livelihood and retirement.

Each calendar year, TCDI makes contributions to the ESOP in cash or company stock. The contributions are invested by the ESOP exclusively for the benefit of each vested employee. The ESOP is designed ultimately to provide beneficial ownership of company stock for employees, the same people who are primarily responsible for the success of TCDI. The longer an employee stays with TCDI, the greater their vested interest is in the plan.

The success of TCDI depends on the teamwork and positive attitude of all its employees. No one has a stronger interest in caring for and promoting the business of TCDI than the people who benefit from its growth.

Trust is a two-way street.

        66 *He who does not trust enough, will not be trusted.*

                              – Lao Tzu

A Collection of People

Our staff at TCDI always have had an admirable work-life balance. It's almost required. James Howell's proverb "All work and no play makes Jack a dull boy" resonates company-wide. We are not defined by our job titles but lead exciting lives outside of work. Being excellent at work spills over into our personal lives.

Or is it the other way around?

Project Director **Jason Bentley** coaches high school girls' soccer at Merrol Hyde Magnet School in Chattanooga, TN. In 2022, he led his team to their second consecutive state title as the Class A Champion. From the *Nashville Tennessean*: "Going back-to-back is not an easy thing and I'm just thankful to be a part of it," Merrol Hyde coach Jason Bentley said. "It's great for the program and great for the players. I think last year winning it, we had to manage the pressure a little different this year whereas, in years before, it was striving to obtain. We focused in on that while last year was great, but we weren't defending that title. We were going out with the goal of winning a title for this team and make that the mindset. The girls focused in on that well."

Jason (far left) and his championship team

5

A ZEBRA IN A HERD OF HORSES

Well before being unique was on trend, TCDI paved its path early on to set itself apart from competitors in the technology industry. Like a black and white zebra in a herd of monochromatic horses, we're unique.

In 1995, R. J. Reynolds became a client of TCDI. The tobacco giant needed a tech company to help them migrate from one version of software to the next. R. J. Reynolds had been rejected by several larger tech companies because of its popularity in the tobacco industry. When the vast syndicate reached out to Bill Johnson, he jumped at the chance to assist. At the time, TCDI only had 10 employees.

Initially, R. J. Reynolds only wanted a three-month contract to expand their litigation systems. Once they realized our unique tech methods and professionalism, the short-term contract grew into a multi-year contract. Almost three decades later, R. J. Reynolds is still a client of TCDI.

We skillfully ease our clients' stressors. We are in the business of damage control and do the heavy lifting for our clients. Whenever a client faces a lawsuit, we're there to help them with their technological needs. We help ease that stress.

We avoid saying "No" with a can-do attitude. We don't get wrapped up in why something can't be done, but in how we can solve problems for clients. We are efficient, have few wasted cycles, and keep looking for ways to do better.

Be different from the rest of the pack. That was always our mantra.

In 2022, TCDI acquired Aon's eDiscovery Practice. The process started in 2019 when the corporate market was high and companies looking to be acquired held the advantage. Like a seller's market when buying a house, a company often chose the highest bidder and procured the most money from the sale. Aon was no different. The financial services firm had a handful of interested buyers, including TCDI. But TCDI was not the highest bidder. TCDI did not want to jeopardize its well-established niche in the market by spending more. TCDI stressed its uniqueness, professionalism, and integrity. Aon had a similar corporate culture and ultimately chose TCDI, forgoing more money from other bidders. Aon's decision proved that more money is not always the best way to go.

During the acquisition interviews, Bill was the only CEO to ask the eDiscovery staff at Aon how they felt about the process and what they would like to see happen. He approached it with excitement like he was helping *them* succeed, not the other way around. Going with TCDI was best for the business with a company that was vested in the longer-term strategy.

Thirty-five staff members from Aon's eDiscovery team joined TCDI in July 2022. The integration was collaborative, welcoming, and engaging, and there were no preconceived notions. Bill was appreciative of how Aon's eDiscovery team ran itself and allowed the new staff to maintain their career paths.

Was there some initial hesitancy? Of course. What company doesn't have that? Fear of being forgotten in the grand scheme of things plagued the new team members. But Bill and the rest of our current staff eased their minds with hospitable professionalism and care. The new staff from Aon slipped right in and has been busy with projects and clients ever since. The colleagues in Aon's eDiscovery practice joined TCDI and will further strengthen and expand its staff knowledge base and experience in the eDiscovery industry.

With the closing of the deal, TCDI added expansive advisory services such as information governance as well as the NOMAD mobile processing platform, PHI/PII detection, and post-data-breach support tools. As described in an earlier chapter, eDiscovery is the electronic aspect of identifying, collecting, and producing

Things can change quickly in the tech industry. We understand that and are lucky to have many clients who are great partners. A client might think they have the best solution, but sometimes we offer them a better option. The best clients listen and take heed.

Mutually respectful clients don't impact projects on holidays or vacations. But we may have to work 24/7 until an issue is resolved. All Project Managers (PMs) have a direct interface with their clients. The PMs want their clients to love everything that we do to make their lives simpler. Clients trust that everything is handled. If we build everything correctly for them, the client can sleep better at night.

electronically stored information in response to a request for production in a lawsuit or investigation.

"We chose Aon's eDiscovery practice not only because of the strategic benefits to our customers but because we share very similar business philosophies and values," said TCDI Founder and CEO Bill Johnson in a press release. "Our companies were both built and grown with client service and continuous process improvement at the core, and our customers can expect the same high quality of service and expert support as we move forward."

Aon and TCDI will maintain a strong commercial relationship to support ongoing and future client engagements and Aon's broader cyber solutions will enhance client delivery and support.

As a Senior Director, Client Services by day, **Sara Coley** thrives on labels, color-coding, and organization. This explains why her coworkers once described her as diligent, responsive, and meticulous. As a yoga teacher by night, Sara lets go of the need for order to go with the flow. She'll answer your call but she may be in tree pose.

Data Mining Manager **Meghan Pertler** is a single mom raising two independent, strong, young women: Mallory and Maggy. When Meghan went through a divorce in s 2009, her biggest fear was that both of her daughters would be affected, so she tried to find something to be a distraction. Horseback riding was the first thing she found and little did she know it would change their lives forever. Both of her girls ride and have had horses over the years. Both are prestigious World Champions, each winning their divisions at different times at the "Super Bowl" of shows: The World Championship Horse Show in Louisville, Kentucky.

Mallory was lucky enough to win when she was only thirteen years old aboard her horse Memphis Gal (Mazy). Maggy won in 2019 when she was 16 riding her horse Fred Lobster (Lobster). Meghan's girls are humble, put in long hours at the barn, and are the hardest workers she knows. For a decade, both have traveled all over with their horses to compete with their barn Krussell Stables in Richmond, IL. The girls maintained a 4.0 GPA all while missing many weeks of school because their show season ran from May through November. They both have sacrificed friendships, sleep, and normal childhood lives to live out their passion for being around horses. Meghan is incredibly proud of both of them.

Horses taught Mallory and Maggy patience and that hard work pays off in the end. Meghan has been their biggest supporter over the years and so thankful horses have come into their lives.

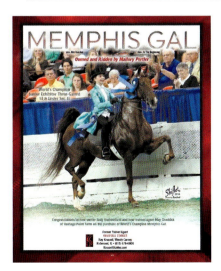

6

SMART IS SUCCESSFUL

Bill Johnson's initial goal was to surround himself with the smartest people he knew. Having a perfect SAT score or being an alumnus from an Ivy League university wasn't the baseline for bringing in talent. Those factors didn't hinder someone, but they weren't the deciding elements either. Our best team members have a combination of street smarts, business smarts, self-confidence, creative abilities, and outside-the-box thinking. Bill looked for folks who had a point of view without being egotistical, and who were willing to learn from other smart people. He wanted likable people to join the company. Existing staff members needed to have a positive reaction to the newcomer. Everyone needed to be cohesive and contribute to the company as a whole to create a viable product.

Thirty-five years and 175 employees later, that mentality still holds true at TCDI. Smart people are good together regardless of their individual talents. We need to think for ourselves and search for answers. We value each other and have the same goal.

We are like master writers who close our eyes and imagine how the final scene will come to fruition. Some of us don't have college degrees and others might not be utilizing the degree we have. But we all aim for the same target.

Bill creates an innovative culture by bringing in some of the best and brightest minds in the industry and fostering an environment of creative freedom. He wants to give us the platform to realize our dreams. If someone had the idea to condense coding steps, they held the floor. If someone else suggested a chili cookoff to increase morale and earn bragging rights for their culinary skills, the event was set up. With the aforementioned round organization chart, we reversed the idea of staff waiting to be told what to do by upper management.

Bill is smart in all fields of study. He reads and studies. He is generally quiet, but can eloquently talk to a room of people with intellect and ease. Some adjectives that describe Bill are: generous, gracious, caring, intelligent, knowledgeable, smart, professional, calm, eloquent, trustworthy, perceptive, and strategic. He provides vision and leadership and is intimately knowledgeable of all company details as well as many applications and software.

Given the initial philosophy to "work with the smartest people," Bill has made employee retention and loyalty a primary value. Employee number four remains on TCDI's team, entering his 30th year at TCDI. Twenty-six employees have over 10 years and more than 60% of staff exceeding five years of service. Bill works to hire the best, provide the most challenging environment, and build a workplace deserving of the finest staff.

Like a master chess player, Bill thinks five moves ahead. We might not know the initial steps to get where he is intending, but we trust him to lead us there. We have faith in him that he knows what he's doing to attain a checkmate.

We all at TCDI have different personality types. Extroverts bring introverts out of their shell. Type Bs tell Type As to relax. We might not forge a direct path from Point A to Point B, but the end goal is the same: find a solution to the problem. This is our aspiration for excellence.

As a minority-owned company with 35 years of experience providing litigation technology software and services to some of the most public litigation in U.S. history, TCDI has learned a thing or two about excellence.

Chief Technology Officer **Chris Attucks** has been coaching high school, AAU, and USA Track & Field since 2010. Several of his state champion sprinters have added to their trophy case with All-American recognition. When he's not coaching track, Chris volunteers at Experiment in Self Reliance, Family Child Services, Urban Ministry, and YMCA.

7

CANINE COMPANIONS

While growing up in Japan, Bill Johnson's family neighbors had dogs. He'd watch them run through the minuscule yards of Tokyo and play with the neighborhood kids. He begged and pleaded with his parents for a furry friend. After a few months of relentless requests, his parents finally acquiesced and brought home a cocker spaniel. But living in a marginally evolving first-world country in the 1960s provided obstacles for dogs. Animal care was not prevalent. The first five Johnson family dogs developed distemper and did not survive after a few months.

Bill's father was a civilian working for the U.S. Army and the family temporarily moved to Syracuse for 18 months. While in New York, the Johnsons welcomed a Maltese and a Shih Tzu into their family. When they returned to Tokyo, the dogs were a perfect size in a crowded country.

As an adult, Bill never went without having a Labradoodle by his side. He decided soon after his stretch at DRC that he wanted to own his own company so that he could create a dog-friendly atmosphere. He could bring his dogs to the office whenever he wanted.

Fred and Ginger lived to be seven and twelve, respectively, and Bill and his wife Susan are the current human companions to Teddy and Chip.

One afternoon in 2008, Susan brought Fred to the office in an SUV. She unexpectedly had to leave and, at the end of the day, Bill had to bring Fred home. The only problem was that Bill drove a two-seater Ferrari with leather seats. He couldn't leave Fred at the office by himself, so he motioned for Fred to jump in the front seat and the two of them took off down the road. Fred's tongue flapped in the wind and his eyes glistened with happiness as they passed other cars and houses whirring by. Bill could have sworn that Fred was waving to all of his doggie friends and bragging, "Look at my ride… You're in a minivan."

Again, Bill saw an issue and solved it.

Bill trusts us at TCDI to dog-sit at the office if he has an offsite meeting. When Fred was a pup full of energy, Bill dropped him off one night at the office when he knew a few employees were working late. Not only did they watch Fred, but they ran him around the hallways and tired him out. Fred slept well that night.

Lately, Teddy and Chip meet staff and guests alike at the front door, with some welcome barks looking for belly rubs. The Labradoodles are TCDI's personal door greeters perched on the sofa in the reception area.

Bill's administrative assistant, Christine, is a dog groomer and dog sitter on the side. When he and Susan go out of town, they often bring their pooches "To the spa at Christine's."

We are encouraged to bring our friendly dogs to work and post cute photos of our precocious pups on team message boards. Some of the dogs search trash cans foraging for discarded snacks. The regulars earned TCDI ID badges attached to their collar. When a staff member had a stressful day, it helped to snuggle and hug the four-footed givers of unconditional love. In 2019, we sent out a holiday card full of photos of staff pooches. The TCDI website even says, "Being a dog lover is a plus."

Joey Adams, Director of Systems Operations, was always interested in woodworking. In 2013, he wanted to expand into metalworking and welding. As his garage filled up with equipment and tools, he felt there had to be a better way that wasn't so cost-prohibitive. He had heard of hackerspaces where artisans could take things apart and put them back together. The collection of metalworkers, carpenters, and sculptors created a community of shared resources and tools to work together and pitch ideas. Joey took over an online Wiki about the community and set up a website and the name Forge Greensboro. Initially, the group met twice a month at a local coffee house to talk shop. Twenty people came. Then thirty. Then forty. The group soon decided that they needed more space and incorporated themselves as a non-profit in North Carolina. Over thirty people signed up to be paying members for the group to rent a space. In 2014, Joey was awarded a *Triad Business Journal* Top 40 under 40 honor. President Obama invited him to the White House to discuss how the federal government supported the alliance. As membership grew, Forge Greensboro acquired more equipment and skilled artisans. The community took notice, too. TCDI is a major sponsor. Forge Greensboro now collaborates with high schools, community colleges, and apprenticeships within the city workforce programs, benefitting the homeless and unemployed. As of 2022, over 215 members contribute to Forge Greensboro. The Forge is a haven for inventors, students, entrepreneurs, artisans, trade professionals, tinkerers, and hobbyists. By mixing ideas, trading perspectives, and sharing skills, the makers build community while they use the Forge. www.forgegreensboro.org

JOEY ADAMS
PRESIDENT

8

A Labradoodle in the Body of a Chihuahua

It's rare for a company of our size to have so many Lean Six Sigma certifications – most companies are much bigger. Lean Six Sigma is a process improvement practice designed to eliminate problems, remove waste and inefficiency, and improve working conditions to provide a better response to customers' needs. Being a Lean Six Sigma company makes us uniquely qualified to help our clients manage their fast-growing litigation and data volumes by utilizing proven waste- and defect-reducing methodologies.

While the development of an innovative culture requires a strong foundation rooted in proven methodologies and theory, it cannot sustain without a top-down investment in the people who do the work. We accomplish this through education and empowerment. Team members are trained through the basics of Lean Six Sigma philosophies. Yellow, Green, and Black Belt certifications were provided through Alamance Community College and North Carolina State University. Since 2013, we have invested over 5500 hours of training and Lean Six Sigma certified process experts account for over 50% of TCDI employees.

By 2019, we had a firm understanding of the impact that LSS had on TCDI in the way we think, act, and approach process efficiency. Six years of LSS training had taught us a lot including that we could make the learning and impact even better. We broke free from the college system and created our own Lean Six Sigma training material and certification process. It's a vigorous program that contains specific applications to our industry and defensibility of process.

TCDI has many employees who have earned belt levels in Lean Six Sigma. We have one Master Black Belt,

13 Black Belts, 16 Green Belts, and 50 Yellow Belts. All have graduated from the Lean Six Sigma Hat Factory over the years where they learned to simplify processes. One year, they were pitted against a group of fifth graders. The exercise illustrated how even smart adults can over-complicate simple processes and be outfoxed by a group of youngsters who see things on a much clearer and simpler level. The competition encompassed an entire business model making three types of hats (sailor, drummer boy, and princess). Much to the chagrin of our TCDI contenders, the fifth graders won because they figured out more efficient ways to make the hats.

Even in defeat, our TCDI staff still learned something fundamental.

In 2013, Geoff McPherson joined TCDI after he met Bill's wife Susan. Geoff had been working with Susan's ex-husband at the time. She recognized the need to bring Lean Six Sigma to TCDI and brought Geoff on as a consultant. Two months later, Geoff found his home at TCDI with spacious offices with beautiful views of the golf course, stocked kitchens, and wonderful, smart people. But the leadership impressed him the most. Bill and Susan actively ensured Lean Six Sigma would infiltrate the way the company thought and acted. Susan was a natural; process efficiency and disgust for waste were in her blood. Self-proclaimed "lazy", she wanted to find the easiest way to do everything. "Lazy" was a far cry from who she really is but she thrived at making things better. Susan was in the first Yellow belt course that Geoff taught and she eventually became certified as a Lean Six Sigma Black Belt. That top-down approach spoke volumes to everyone at TCDI.

What does it mean to have Lean Six Sigma Certifications?

TCDI proves its worth to clients by providing a solid way to solve problems or reduce the impact of major changes. For example, we have provided numerous new clients a way to 'clean house' before making any significant data moves. We use the 5S system, a core tool of Lean Six Sigma. The 5Ss are for organizing spaces so work can be performed efficiently and effectively. This system focuses on removing clutter, putting everything where it belongs, and keeping the workspace clean, which makes it easier for people to do their job without wasting time. It raises product quality and improves work productivity, resulting in lower costs and higher efficiencies.

If timing permits, utilizing the 5Ss before making such a huge undertaking as a data migration can save money and time. It also allows for a clean start with a new vendor. If the timetable doesn't allow streamlining before data migration, cleaning up the data must be one of the top priorities immediately after the dust settles and before the status quo sets in. Here is a brief overview of the 5S process:

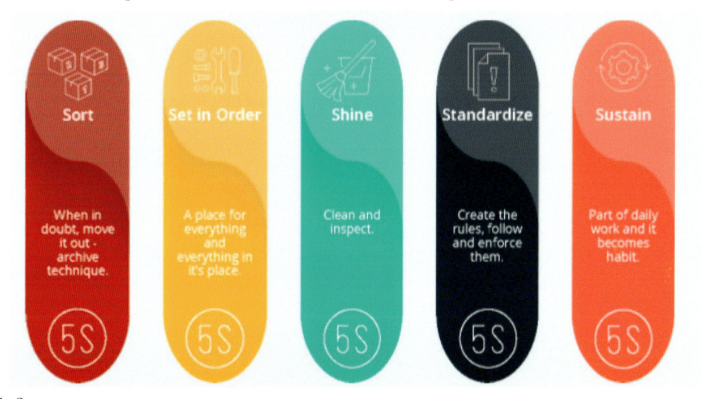

1. Sort

This is often known as cataloging in the data world. When in doubt move it out, which could involve archiving databases, hardware, records, fields, review templates, or even users. The objective is to identify everything that can be removed or no longer needs to be retained, which in turn will help to remove clutter from the new working environment. Start by asking the following questions:

- What is the purpose of this data, file, or document?
- Who uses it and for what?
- How frequently is it used?
- When was it last used?

2. Set in Order

During this process, there is a place for everything and everything is in its place. For example, the objective could be to evaluate the current standard database structure and use the migration as an opportunity to create a working environment that is sustainable for current and future litigation needs. Start by asking the following questions:

- Who uses what data?
- When is the data used?
- Which data is used most frequently?
- Where would it be most logical to place the data?
- Would some placements be more accessible for people than others?
- Would some placements reduce unnecessary searching?
- Are storage containers or directories necessary to keep things organized?
- Are all the fields needed?

3. Shine

With the clutter gone and the storage space organized, it's time to clean. After a thorough initial cleaning, staff tidies up their projects every day. This maintains the gains made in the Sort and Set phases. The Shine phase creates a more pleasant environment for employees, who no longer have to combat a cluttered database. Cleaning the area every day also leads to a higher level of employee buy-in for the 5S method.

4. Standardize

Create the rules, follow, and enforce them. Standardize is all about maintaining the new eDiscovery structure. The objective is to create the rules for setting up new databases/workspaces, fields, indexes, users with administrative permissions, etc. Document them and build governance for adherence. Sloppiness in the area of standardization and documentation can lead to the unraveling of a lot of hard work.

5. Sustain

This is a key part of daily work and that should become a habit with standardization and consistency. Sustain is about making the 5Ss part of the culture. Things stay a particular way because people know it as the norm. However, we are talking about people – humans. Things tend to get out of control over time and if not implemented properly can find having to apply the 5Ss repeatedly. This process is only as good as all of the 5Ss are implemented. The objective is to develop and provide training for everyone who will interact with the new protocol. Develop an audit schedule to oversee key processes to ensure they are compliant. Create a process for exceptions. Be prepared for deviations. An escalation path for this needs to be defined in advance.

Transitioning data and workflows to a new vendor can be made easier with the right tools and partners. The 5S methodology is one of many ways to make a vendor migration less stressful, better organized, and well-documented.

A closet Excel nerd, **April Marty** is multifaceted and constantly uses both sides of her brain. One minute, she's designing marketing collateral, and the next she's creating in-depth sales reports. She can carry on a conversation about PMS colors that flow right into pivot tables and VBA. But don't ask her about the weather. April is not a fan of small talk. Things she likes? French fries, karaoke, *Saturday Night Live*, and iced coffee. How else do you think she gets it all done?

9

A CEO IN FLIPFLOPS

Many other CEOs prefer to have their formal portraits prominently displayed on the walls of their corporate headquarters. Not Bill. He finds getting his picture taken a necessary evil. His abhorrence of the camera came when he was growing up in Japan. Bill's mother approached a Japanese company and convinced them that young Bill and his brother Charles would be perfect faces for their commercials. The company produced a Nutella-like sandwich spread that, according to Bill, tasted like a bitter salty concoction that should not have been force-fed to anyone, let alone a couple of kids. The advertising execs encouraged Bill and Charles to plaster smiles on their faces throughout many commercial takes. The fallout from that unpleasant day still makes Bill associate cameras with something foul.

Bill is not a typical CEO. He has created a company and culture that is unique to his persona. It's common to find him standing by the grill cooking lunch for all 175 of us or staying up throughout the night to help meet a client deadline. He leads by example and never asks anything of us that he is not willing to do himself. He is an outlier compared to most other CEOs.

A great leader shows sincere enthusiasm for the business, its products, and its mission. Bill holds himself to the highest level of excellence and expects nothing less from us. He works alongside us pulling all-nighters and addressing issues that arise. He stays ahead of industry trends and his expertise challenges and sharpens our skills. Whether acknowledging his mistakes, giving credit for accomplishments, or putting quality above the bottom line, Bill exhibits integrity in all he does.

Transparency is important to Bill. He openly discusses high-level business plans with our entire company through regularly scheduled town hall/all-staff meetings. He sets the tone from the top, hires selectively, recognizes and rewards successes, and celebrates wins. He fosters open communication and dialogue throughout the organization regardless of title or tenure.

Bill is especially sensitive to work/life balance. He likes to ensure we are rewarded for the time we spend away from our families. Bi-annual company events are planned to include spouses and children to honor the support the family gives to all employees.

The culture at TCDI reflects its keys to success from its casual, yet involved, CEO.

Known as a team player and a knowledgeable asset, **Melissa York** has a knack for cleaning up "messy" data and processes. It's no surprise that her favorite show as a child was *Double Dare* on Nickelodeon. She still has one of her first projects, the dress she made in 12th grade Home Economics. Melissa has worked with TCDI since 2006 as an employee and contractor. She has used her legendary communications skills and attention to detail to liaise between clients, internal teams, employees and vendors in her project support and human resources admin roles.

Since 2003, **Ned Adams** has been providing unmatched service to TCDI clients. The only thing he's done longer is support his UNC Tar Heels, and rock the Birkenstocks he still has from college. Ned is known for showing up to work early, like 5 a.m. early. Maybe that's why he is so beloved by his clients and colleagues alike. Ned works to the beat of the *Magnum P.I.* theme song. His secret could also be his dedication, communication skills, and dependability. Either way, this Director, Client Services is most likely planning for large, complex litigation matters, or potential golf trips he'll never go on.

Ned and his daughter Avery

10

TAKE CARE OF YOUR PEOPLE

Bill Johnson believes in going above and beyond for his staff. They are, after all, the lifeblood of his company. He wants them to feel at home–and they do. TCDI offers a competitive salary, unlimited time off, fully paid healthcare for employees, and HSA reimbursements. More tangible benefits include a stocked kitchen, an on-site gym, access to an in-house trainer, and a Pac-Man game.

Annual chili cook-offs are a point of pride at TCDI, especially in the heart of North Carolina barbecue country. We compare recipes and hint at secret ingredients, giving us a chance to talk about something other than work. Every December, we contribute to an annual holiday cookbook full of recipes for pineapple cheese balls, gingerbread hot chocolate, French coconut pie, and Jack Daniel's beef tenderloin.

We do unsurmountable work and get equally rewarded for it. At holiday parties, Bill may hand out TVs, game systems, headphones, or luggage, while other companies randomly pass out $25 gift cards. One year, Bill acquired tickets to the NCAA Final Four in New Orleans. Ohio State was going up against powerhouses Kansas, Kentucky, and Louisville then. He knew that one of his employees was an Ohio State fan and invited him to tag along and cheer on his favorite team. Bill has also gifted tickets to the Orange Bowl in Miami. During the annual holiday party at Sedgefield Country Club in Greensboro, employees played Family Feud and won gift certificates, cash, and lottery tickets.

Bill was selfless enough during TCDI's early days to deny giving himself a paycheck. He recognized the time delay that occurred from TCDI being paid by the slower government agency processes. Rather than letting

his employees suffer, Bill forewent drawing a proper paycheck. He made sure his people were paid first. He took care of us before himself. Because that's what kind of person he is.

New staff sometimes say, "I'm still trying to figure out when the other shoe will drop." They aren't used to coworkers being so helpful, kind, and accommodating. We do more than focus on salary and benefits. In an industry where tech employees job-hop, we *want* to stay.

Having fun is encouraged at TCDI. Before most of our staff worked remotely, we used to set up a haunted house at the office every October for everyone to enjoy. The first floor of headquarters was transformed from a stark office space into a dark, ghoulish atmosphere in 48 hours. We dressed up as witches and ogres to participate in a costume contest, hoping to win prizes and bragging rights.

On birthdays, the celebrant would be susceptible to pranks like having their office filled with three feet deep balloons or strung with yards of toilet paper. One staff member found his telephone strung up the wall, in the ceiling tiles on his big day and ringing 100 feet away down the hall.

Another unsuspecting employee handed his keys over to his coworkers thinking he was getting a car detailing for his birthday. He trusted them; they were like family to him and TCDI paid for the service. Why would he think anything differently? An hour later, he came face to face with "Honk if you love Taylor Swift" on his back window. Pink and white balloons filled the inside of his car. On the front windshield, his prankster coworkers had written, "Hey y'all, it's my birthday!" and drew a frame around the driver's side. He had to drive 90 miles home that day with passing cars honking at him the whole way.

Another employee had his office turned into a cave. His family owned property near a cavern and gave tours. When he returned to his office, he had to get on his hands and knees to crawl into a makeshift hollow full of cardboard stalagmites and stalactites. Glow sticks lit a path through the blacked-out room. His prize for being a good sport: an Indiana Jones hat and whip.

One of the new eDiscovery team members from the Aon merger learned early on about Bill's generosity. He found out she loved NASCAR and arranged for her to meet American racecar driver Kevin Harvick.

We are expected to grow and show our personalities. When the weather is nice, staff have been found outside on the parking lot after hours tossing up some jump shots (and maybe some bricks) during a pickup basketball game.

If Bill isn't in his office, there is a good chance he is out back grilling tuna steaks for us, or crashing a meeting to collaborate on new features or client requests. He's been known to have casual meetings out on the patio at TCDI headquarters (even in the winter with portable heaters) to make us feel comfortable around him.

Not only do we build great relationships with our clients, but we also have strong bonds with each other. With an average tenure of 13 years across the company, we are comprised of data scientists, developers, attorneys, professional client service, and subject matter experts in the field.

Skill metrics are kept on each of us and quarterly evaluations are done. This is not a way to look for staff faults, but instead to increase business value and skills. On occasion, an employee may not be performing up to par. A senior manager may step in and say, "Give me six months to improve their performance and save their job." The manager will set up additional training and skills testing and the employee might not even know the logic behind it. Most companies won't take the time or effort to assist their staff and many companies end up letting the underperforming staff member go.

In 2022, our exceptional work was noticed. A long-time client of TCDI extended a project to a subsidiary, knowing that we could do stellar work. For months, the TCDI sales team asked for feedback from the subsidiary but to no avail. The sales staff forged on because they had a deadline to meet. When the project was complete, the subsidiary didn't like the final project and only wanted to pay 1/6 of the invoice. Bill stepped in and said, "No, we won't reduce the amount due. My team is worth the full invoice. I won't have you devalue my team. But what I will do is void the entire invoice because we no longer want to do business with you." Bill lost business with the subsidiary that day but knew that he had our back. The long-time client took notice and supported Bill's decision.

Bill Johnson takes care of us. In an industry where the norm is to bounce from company to company every few years, we stay at TCDI for the long haul. It's not simply a paycheck, but a family.

As Vice President of Legal Services, **Michael Gibeault** is often traveling to meet with clients and prospective clients. It makes sense that, given the choice, he would choose time travel as his superpower. Michael is ambitious, energetic, and personable, which is why he excels at building client relationships. Or are they only impressed with his shoes? He's a self-proclaimed shoe snob. Does your dream-shopping spree include Salvatore Ferragamo? Michael's does!

11

TAKE CARE OF THE CLIENTS

By Dave York, Managing Director, Litigation Services, and Geoff McPherson, Chief Process Improvement Officer

Over the years, we have done many joint presentations for clients and co-workers alike, but there is one in particular that stands out most in our memory. It doesn't resonate because it was our best or most insightful, but rather because it was arguably our worst. It is a client presentation that we refer to as the "Red Line, Blue Line" presentation, and it was only saved thanks to Bill Johnson.

A meeting had been scheduled with one of our largest clients and we were to apply TCDI's newly adopted Lean Six Sigma (LSS) principles and present to the client our application of the LSS methodology to solve a key data challenge. The only problem was our measurements and analytics were not quite giving us the results that we had hoped for, and a key part of the project's success hinged on software development work that had not happened, nor was it in the immediate pipeline. We conveyed our concerns to Bill and desperately wanted to delay the meeting, but it was too late to reschedule.

The day before the meeting, we stayed at the office late to work on the presentation. That evening, we both stressed and struggled to sleep with the pending dread of what was sure to be a disappointing presentation. Throughout it all, Bill remained his usual calm self. He showed up at the airport more refreshed than we were and he was more relaxed about the meeting than we were.

We arrived at the client site early. Instead of going straight to the client's office, Bill took us to Arby's for a pre-meeting meal. We ate our beef and cheddars in silence and drove to the client's campus. When we reached the office, Bill pulled over and let us out on the sidewalk, and said that he would park the car. Both of us got out and looked at each other. "He's not coming back, is he?" Geoff said. We were jokingly convinced that Bill had fed us our last meal at Arby's and then dropped us off to send us on our way. Luckily for us, Bill did come back.

Once inside and presenting, we went through our presentation slides with the audience. We concluded with two slides that showed before and after results using a graph with a red line and blue line to show pre- and post-process improvements. We can't remember which color represented which metric, but the general premise was that if we did X, Y, and Z, then the red line value will increase above the blue line value, thus improving the overall process. The only problem was, we had no immediate way to implement the proposed solution, nor did we have a timeline for when we could do so. After a few seconds of silence that felt like hours, this is where Bill stepped in to save the day.

Bill had a clear vision of what needed to be done by our Development team. He set forth not only the proposed solution but the timing of its completion. He didn't do it with a presentation or handouts, and not a single red line or blue line. He just used his words, but he did so in a way that conveyed confidence to the client that their problem would be solved.

Confidence, caring, and commitment summarizes the story of every client meeting that we've ever been in with Bill. He won't talk the client's ear off, but when he does speak, clients listen, and they realize that there is a top-down commitment at TCDI to solving client problems.

Bill and Susan have always been committed to and encouraged client meetings and visits. They are as close to some clients as they are to their employees and family. Their presence with existing and potential clients can turn challenges into opportunities. It's not uncommon for Bill to take on client challenges that other vendors shy away from and to do so successfully on a regular basis. Much like Bill's words in client meetings, the strategies and decisions that are methodically made are driven by a key initiative to take care of the people at TCDI and take care of our clients.

12

INTEGRITY ABOVE ALL ELSE

TCDI has never been a screaming, in-your-face company where the lowest prices support the bottom line. A reverse auction, where the contract goes to the seller willing to accept the lowest amount, would never be successful for us. We refuse to sacrifice quality and integrity just because a prospective client requests a project done cheaply and quickly.

We've forged success with clients by being honest and true. Having the lowest price was not our driving factor. We lost clients by not being the most cost-effective. In the long run, we realized we didn't have to take on every single client to be successful. Our integrity took the forefront and we have used that as our benchmark.

Focusing on quality and relationships has always been the most important pillar of success at TCDI.

Our account managers get to know their clients and foster deep relationships with them. They might ask the clients about their family, recent vacations, birthdays, or books they've read. If a client emails various questions to TCDI, the account manager may respond with, "Can I call you?" and then they have a conversation instead of a sterile email exchange. Personal relationships matter.

We understand our clients' goals and who they are. We speak in plain language to non-tech clients and translate their legal needs into practical solutions. This innovative style worked with a lot of people who were not technologically savvy. Then, instead of acting independently, we asked for feedback from our clients. Going above and beyond is expected, not an exception. Listening to clients' needs is at the forefront at TCDI, even if it means encountering a detriment with our products or service.

Other pillars of success at TCDI include:

- Tell the truth

- Encourage positivity

- Respect everyone

- Give credit where it is due

- Collaborate, don't compete

- Value diversity

Building long-term partnerships with clients have resulted in client relationships extending beyond 20 years, and employee tenures that are equally as long. Our philosophy ensures that design engineers and data scientists have a true understanding of clients' obstacles and challenges, which allows us to adapt our solutions to meet the clients' needs. This eliminates the burden of working with third-party development teams, resulting in increased efficiencies, and savings for our clients. Bill has always said, "Once a team is built with clients, we believe that team needs to stay together," and this shows in the relationship-centric culture that we have established.

Customers tell us that we feel like family and a true extension of their teams. That is deeply gratifying as we have spent the past three decades fostering a people-

" *The way a team plays as a whole determines its success. You may have the greatest bunch of individual stars in the world, but if they don't play together, the club won't be worth a dime.*

– *Babe Ruth*

Altria – a tobacco company with 6000 employees headquartered in Richmond, VA, with a revenue of $21 billion

"TCDI has been Altria's go-to eDiscovery provider for decades, and what sets TCDI apart from its competitors is Bill Johnson's leadership. Bill's vision is simple: do well by doing good. Only a leader believing he could build his business by first serving members of our armed forces could have imagined a Military Spouse Managed Review program. Now, no matter the remoteness of their post or station, military spouses are meaningfully employed using TCDI's world-class technology and project management to conduct remote document review. Remote review may seem normal now, but it was unheard of at the time Bill launched TCDI's program."

Kim Harlowe, Senior Director Litigation Support & Technology, Altria centric culture focused on providing excellent customer service. Our goal has always been to cultivate long-lasting relationships and design all processes with that goal in mind.

Clients engage with us because something bad has happened. They are often being sued and need to retain sensitive documents for an extensive and expensive trial. They now have to spend money on something other than what they are in the business to do. They need to trust a company to do a job and take care of them. That's where we come in.

Integrity.

That's how we have been able to retain many clients for an average of 10-15 years.

As the third of four boys, **Dave York** learned several valuable life skills early on, including cat-like reflexes from impending midsection punches, how to test a 9-volt battery with his tongue (something every kid should know), and how to work hard, enjoy family, and laugh as much as you can. Dave was voted most likely to start a game show by his coworkers. His strong work ethic has benefited not only his career but also his fundraising efforts with the Leukemia & Lymphoma Society (LLS). In 2018, he even climbed and summited Mt. Kilimanjaro with his brothers to raise money for LLS!

Karlotta Young, a Senior Accountant in Finance, has lots of hobbies. To begin with, she's an avid runner who logs three miles a day five days a week. For the past 10 years, she has taught yoga and Pilates at the Burlington, NC, YMCA. Before she took it over, the class was choreographed by former ballerina Dawn Lombardi for 45 years. Karlotta also plays the flute and is aspiring to play with the Worship Team at her church. Lastly, since she was born and raised right outside of Pittsburgh, Karlotta is a passionate Steelers football fan.

13

SUSAN BRIGHT - AN INSTRUMENTAL ASSET

By Bill Johnson, CEO

There are many ways to reflect on the history of TCDI, but one of the most interesting is to map the history of TCDI to that of a person moving through the various stages of human development. Following its birth in Northern Virginia, TCDI entered its teenage years as it evolved from a typical Federal contractor into a firm specializing in litigation software and support. For many a person, when looking backward through the formative years, some of the most critical influences which lead to a life well lived are gathered during these teenage years. There are many stories of that one teacher who helped set the stage for adulthood with lifelong lessons. In the case of TCDI, that person of influence appeared in the form of Susan Bright.

Susan entered the TCDI universe through a law firm client, where she was working one of her three jobs as the Director of Computer Projects within the law firm. If you asked Susan why three jobs, her answer was that she never wanted to feel beholden to any single job and that the ability to maintain her personal integrity if she felt a job violated her personal ethics by having the freedom to leave was important, and not having any single job become so critical to economic survival gave her a sense of peace and freedom. She had a project management background, but she was also a teacher of computer science at the local community college… actually two of them. With an advanced degree in teaching, she was a natural to help TCDI grow up, helping it learn all of the skills needed in terms of customer service, management, finances, and generally how to be a serious adult.

If you were to create a map of the early leaders of TCDI, you would find that unlike Kevin Bacon and the six degrees of separation, there are at most two degrees of separation between Susan the critical leaders set the path for TCDI to become one of the leaders of litigation support.

For example:

Lisa Cain (CFO) – Lisa is a friend of Susan who knew Susan through her work with the community colleges, having met Lisa during her years at Pfeiffer University.

Eric Ridge (Father of CVLynx) – Susan hired Eric to work at the law firm as a web developer, having recognized those traits in a teenager who cried out "technical genius."

Linda Coleman (Original QC Director) – Susan has a knack for finding talent outside the normal avenues of recruitment, running into Linda as one of her students in computer science with a yearning for dealing with details.

Tad Little (Security Director) – A friend of Eric, so this qualifies as two degrees…or is it one? Tad, whose claim to fame in Susan's book was that of having one of the worst interviews, nevertheless piqued Susan with his knowledge and technical intelligence.

Geoff MacPherson (Chief Process Officer) – When Susan felt TCDI needed a disciplined approach to how it created and executed the processes involved in eDiscovery, she turned to Geoff, first to teach the company Lean Six Sigma, but more importantly convinced him that it was his opportunity to prove the axiom "Those who can, do; Those who can't teach" to be completely false.

The list is long.

Susan took on the role of Teacher in Chief. Her goal of learning how to grow a company, develop the highest levels of ability, integrity, and knowledge, and instill those lessons upon TCDI led her to constantly be in a teaching and mentorship role.

Susan created several initiatives to grow everyone at TCDI, taking it upon herself to give lessons in how to read financial reports and deal with financial reporting to all levels of management so they would understand what the numbers mean. She arranged for seminars on how to have respectful confrontations in a workplace, teaching a group of introverts how to effectively communicate with their colleagues. She led the initiative to bring Lean Six Sigma in-house and to get the technical staff trained. She was part of the first class in LSS to make it through to black belt. When she found out people had an interest in flying lessons, she also organized lessons, including a ground school--nothing that has a specific tie into the work of TCDI but has everything to do with having the right work-life balance.

She was also the principal designer for the TCDI workspace, commissioning local artists to create art for the offices, and working with the architects and designers as we grew from a single-floor office to a three-story building, and into a fourth floor.

So, if you asked the leadership group at TCDI who has influenced their careers and contributed the most to their knowledge, Susan would most likely top most lists.

If you asked our long-term clients who they think of when they think of TCDI, Susan would top that list as well.

And yes, when Susan joined TCDI, she quit all of her other jobs--which may be the greatest endorsement TCDI has ever received. Although a lot of the stories in this book don't specifically reference Susan, she is in every story of TCDI ever since we moved to North Carolina.

14

FORGING A BRAND-NEW PATH

Before the COVID pandemic hit in 2020 forcing most of the American workforce into home offices, we devised a barely-used strategy in 2017: a flexible work schedule based on what millennials wanted.

The average eDiscovery document reviewer was fresh out of law school and under the age of 30. They grew up on sophisticated computer games like Minecraft and Toontown, not the antiquated antics of Donkey Kong and Ms. Pac-Man. Engaging games kept the happy users playing in the comfort of their own homes, often into all hours of the night.

When these young adults started working at eDiscovery review centers, they grumbled at the cold, sterile rooms, three feet of personal space, limited working hours, and sitting amongst strangers. Most centers operated on an 8:00 a.m. to 7:00 p.m. Monday through Friday work schedule.

We recognized this issue and came up with a solution. Bill Johnson welcomed this idea with open arms. He said, "Let's give it a year. If it doesn't work, we'll go to Plan B."

After a few months of researching what millennials liked, how they acted, and wanted to be treated, we created a 24/7 available workday for online review centers. Early objections questioned the technology and security of something like this. Despite that, the young document reviewers were thrilled they could work from home. If they were night owls, they could log on at 10:00 at night, and relax on their couch. If a reviewer had kids, the new online review center allowed them to split their time throughout the day to take their child to lacrosse practice or dance class. Document reviewers didn't even mind submitting photos of their home office for security purposes because they knew the result would be beneficial to them.

This kind of flexibility was unheard of in 2017. The newfound solution worked!

With this unexpected success, we now had to come up with different ways to communicate and monitor the virtual document review center. Reviewers' local machines were disabled from printing, accessing USB devices, moving files outside of the virtual machine, or accessing external email. Review Managers monitor reviewers' activity, conduct training, communicate with individuals or the group, take control of reviewers' screens, and log sessions.

Being remote allowed us to recruit reviewers nationwide. Qualified applicants responded in droves. The working model changed and our clients were thrilled because they no longer had to pay budget-killing overtime. New marketing materials explaining success stories had to be created. Not only that, document reviewers now had an excellent work-life balance and didn't mind going through pages and pages of eDiscovery. Having a flexible work schedule changed their lives.

When COVID hit and sent everyone home, we were ready.

Caragh Landry attributes her long career in eDiscovery to her ability to use common sense and good humor. We think her work ethic, leadership, and ingenuity may have something to do with it! Forever a Project Manager at heart, and a tried and true Scorpio, in her role as Chief Legal Process Officer, Caragh loves to create solutions and lead the way to great outcomes. She expects people to be genuine, authentic, and team players and holds herself to the same standard. Caragh enjoys the outdoors or sitting in her reading chair with a blanket and a good book. When she's in the kitchen, Caragh makes a family-favorite Irish Brown Soda Bread.

15

UNDER THE RADAR

Early on, we garnered clients by word of mouth. Our quiet reputation for professionalism and a quality product spoke volumes. We didn't have to do direct marketing with airport billboards or radio commercials. Word got around the legal tech industry that our quality software and rapport-building were at the forefront. Having referral clients built the business on a small number of long-term relationships.

We had a low profile and liked being under the radar. Our clients didn't want it to be public knowledge that they were in litigation. They had a reputation to uphold. Fortunately, the sales cycle at TCDI was generated from one client ending litigation and another starting. We had a constant flow of one contract ending and another beginning. That rotation was used successfully for years.

Best practices and industry standards existed. We left no stone unturned and went above and beyond what the competition was doing. Being mindful of the confidentiality of clients, we had security and privacy-related controls in place. Team members evaluated additional requirements and certifications before they became best practices. Again, we forged our own path.

Altria, one of our long-term clients, preferred having our undivided attention. Conversely, Altria wanted to sing our praises. In effect, Altria spread the word as external salespeople for us in the legal tech industry. Introductions to potential clients were the lifeblood. This method was a boon because our crux was sales and marketing. With one sales rep, one marketing manager, and a disarray of product names, we knew our weaknesses and we wanted to do better.

But, like anything, the industry started changing. We had to adjust. The marketing team grew from one to four from 2016 to 2018. Additionally, the sales team expanded to 11 and stretched from the East Coast to the West Coast to increase the number of clients. A new branding project set all of the products into a seamless and recognizable suite: CVLynx, CVFox, and CVOnyx. The website, logo, and documentation all had to be redone. A social media presence and brand recognition were born. The effort was Herculean but worth it.

As the TCDI brand became more recognizable, the number of clients has grown exponentially. Each one of the clients remains with us because of the seed of the initial relationship.

Take care of relationships and everything else takes care of itself.

Known as the fastest walker at TCDI, **Anthony Klier** can get from one end of the building to the next in no time. Since 1992, Anthony has impressed us with his integrity, teamwork, and organizational skills. He started writing video games at the age of 13 on a TI-99/4A computer (circa 1983). He also owns a Fender Strat guitar and plays a mean version of "Smoke on the Water."

Not afraid of accepting a challenge, and he accepts a lot of them, **Dave Martyn** is known for his determination and ability to do whatever it takes… and his love of Taylor Swift's music and the Georgia Bulldogs. Since 2011, he has operated with the same energy and enthusiasm he started with. Some say he's a robot, and if he is, we hope he stays powered up for at least another decade!

His Work Mantra: "Go big or go home."

Dave and his stepdaughter Shelby

16
MILITARY SPOUSE MANAGED REVIEW

What's it like to be a military spouse with relentless anxiety about where your next job will be when you are constantly moving around?

Ask Jennifer Andres, our MSMR Senior Director. In 2013, she was newly married and living in Austin, Texas. Her husband David, a Sergeant in the U.S. Army, received orders to report to Fort Hood, an hour away. With a law degree, Jennifer was on a solid career path before she came to TCDI, but didn't have a lot of experience or a traditional attorney salary yet. Traveling an hour each way to Fort Hood for work made her question if her job as a legal document reviewer there was worth it. The time and money she invested didn't equal enough job satisfaction.

Jennifer pondered what would happen the next time she and her husband moved. Would she find a new job? She didn't want to take the bar exam again in another state if she didn't have to. Nor did she want to start with an entry-level position again. The thought of launching the process all over again with every relocation stressed her out.

She knew there had to be more military spouses like her.

In the United States as of 2020, there are approximately 1,200 military spouse licensed attorneys. Despite impressive qualifications and experience, many face challenges to employment with frequent relocation, inconsistent state licensing requirements, and are stationed far from most traditional legal career opportunities.

In 2017, Jennifer joined TCDI. She approached Bill Johnson with the concern that, although she loved her job, she was afraid she would have to give it up when her husband was deployed again. He told her to investigate the need. She posted her idea of 100% remote working on the Military Spouse Juris Doctorate Network and waited for responses. No other company had a program like this and working from home at the time was barely on the radar for most people. To her surprise and delight, other spouses were in similar situations with the same anxiety. She found that even low-paying coding projects were wanted—anything in the legal industry to maintain constant employment.

While her husband was deployed to Germany for nine months, Jennifer temporarily moved to Greensboro, NC, to work at TCDI headquarters. In less than a year of employment, she collaborated with developers, project managers, and quality control to kick off the Military Spouse Managed Review program. The team set up access to a large, nationwide talent pool of qualified lawyers with diverse industry experience and specializations. The virtual team environment fostered collaboration and quality. Our flagship review platform provided a user experience that made reviewers more efficient, with advanced analytics and visualization features making review more engaging and productive.

The project was a success! Clients loved that it fit their diversity initiatives by increasing opportunities for women.

Experienced document review and coders included former Assistant District Attorneys, JAG Corps, law firm associates, in-house counsel, former judges, and solo practitioners. Premier partners included Ford, Shook Hardy & Bacon, Altria, Cumberland Farms, Foot Locker, Whole Foods, Fox Rothschild LLP, and Dell to name a few. Military Spouse partnerships are recognized by the Department of Defense and committed in practice to supporting the military. Over 100 reviewers are fluent in more than 37 languages and dialects.

Jennifer and her husband David

Jennifer could now exhale. She could work at night or on the weekend if she encountered conflicts during the day. The Military Spouse Managed Review program made this happen. For her and other military spouses with a similar career path.

The first week of May is Military Spouse Appreciation Week--a week that celebrates the sacrifices of U.S. military spouses across the country and stationed around the world. This is an important week for us to recognize members of the Military Spouse Managed Review Program which has been going strong since 2018. It is a time for us to honor and thank those who work alongside staff, supporting clients – all the while carrying their families and making sacrifices in service to the nation and national defense. MSMR is a signature program, in which every member of TCDI and client partner takes pride – and one which founder, Bill Johnson, has stated many times is the single most worthwhile endeavor TCDI has ever undertaken.

In 2020, MSMR was already in full swing with a vast roster of military affiliates who are lawyers or have legal experience. Our document review team's job and joy are to engage new military spouses through great organizations like the Military Spouse JD Network, the U.S. Chamber of Commerce's Hiring Our Heroes Program, and the DOD's Military Spouse Employment Partnership, among others. The team strives to provide the most flexible remote work opportunities possible to its teams, knowing well the challenges of military family life, which include frequent moves, licensing hurdles, single-parenting, and providing care for children and parents while attending to home life obligations. Our strong security protocols and pre-pandemic practices which enabled work-from-home long before COVID, are tailor-made for military spouse engagement regardless of where in the U.S. they are stationed. Our Project Managers are well-versed in the demands of military life and they work tirelessly to ensure open lines of communication and maximum flexibility when home and military demands require that team members work odd hours–or even take breaks for yet another PCS. Managers work with attorneys on bases from New York to Texas to Hawaii–all on the same day and on the same project–and provide top work product when the matter is done.

Routinely, at any given moment, military spouses and affiliates make up at least one-third of our contract workforce – most of whom are women – and they are well-represented in every project. A point of pride is our track record and ongoing efforts to identify military spouses who wish to transition to full-time employment – and bring them fully into the TCDI family. We are thankful daily for the talent and leadership that those folks provide and the unique perspectives they offer. We celebrate with contract alumni who have gone on to secure government, academic, or other full-time positions. Whether a military spouse is on staff for a season or many years, we are grateful for their efforts--by showing support for their personal and professional choices.

Not only is supporting the nation's military spouses—and by extension, their families—an incredibly important corporate endeavor, we reap untold benefits by elevating and harnessing the skills, smarts, and talents of a frequently underemployed population. We strive to provide rewarding professional opportunities to folks who are trained to be lawyers, calling on them (regardless of where in the U.S. they live) to use those talents on behalf of clients. It is a story of mutual benefit and success. From their adaptability, flexibility, autonomy, camaraderie, reliability, intelligence, confidentiality, and humor – military spouses are the real deal and the whole package.

MSMR Highlights since inception

- 2000 – TCDI received the Freedom Award–the highest recognition given by the U.S. Government/ Department of Defense to employers for their support of their employees who serve in the Guard and Reserve. Only 15 companies are selected annually.

- January 2018 – TCDI launched the MSMR (Military Spouse Managed Review) program with only a handful of team members. Today, the program is supported by 20 full-time staff and a large pool of project-based contractors, hundreds of whom are military spouses and affiliates – and all of whom work remotely.

- 2019 and 2022 – TCDI sponsored and attended the Veterans Legal Career Fair (hosted by the Orrick law firm) in Washington, DC.

- Since 2020, TCDI has sponsored MSJDN's (Military Spouse JD Network) annual conference which provides programming specifically geared toward empowering military spouses in their non-traditional legal paths and lobbying to ensure state reciprocity in Bar licensing.

- 2020 – TCDI was inducted as a partner into the Department of Defense's Military Spouse Employment Partnership. TCDI reports monthly about military hiring statistics and participates in MSEP Annual Meetings to liaise with other business partners on issues related to recruiting and retaining military talent. MSEP is also a great resource for advertising jobs in hundreds of military outlets across the country.

- Within TCDI's Legal Document Review business line, 10 military spouse contractors have transitioned to full-time employment since 2020, one of whom was made a Director (Jennifer Andres). Each month, at least one-third of all review document projects are staffed with military spouses and affiliates.

- February 2021 – TCDI was recognized by the *National Law Journal* as a Legal Trailblazer for the MSMR program.

- April 2022 – The MSMR Program was named a finalist in the Litigation & Discovery category at the 2022 *Legalweek* Leaders in Tech Law Awards (at the same time, CEO Bill Johnson received the Lifetime Achievement Award).

- October 2022 – The Director of Legal Process at TCDI spoke as a panelist at the Veteran's Bridge Home-sponsored Veterans Hiring Initiative luncheon in High Point, NC, on veteran-friendly employer initiatives along with speakers from Truist and Atrium Health.

- With our recent acquisition of a team from the Aon Group in November 2022, we formally kicked off Data Breach response work which gave us an even greater opportunity to engage (remote) military spouses, veterans, and reservists anywhere in the U.S. Currently, we engage roughly 75 military spouses and affiliates for this work and we anticipate onboarding many more in the coming months.

- TCDI routinely partners with the U.S. Chamber of Commerce's Hiring Our Heroes program, whose leadership refers candidates to them.

- September 2022 – TCDI attended its first on-base hiring fair at Camp Lejeune as a part of their recent Hiring Expo and onboarded several candidates from that endeavor.

• Each Military Spouse Appreciation Day – TCDI sends company swag and gift cards to its military affiliates.

• Every other November, TCDI hosts its annual client induction ceremony via Zoom to recognize its partnership in supporting our Military Spouse Managed Review program.

• TCDI social media platforms are filled with support for military endeavors and sharing information about our MSMR program.

As a Hiring Our Heroes partner, TCDI is committed to providing both contract and full-time work opportunities to 100+ military spouses. Through various military bases, job fairs, and online channels, we successfully recruit year-round attorney military spouse candidates who are seeking online legal work that they can manage from any military outpost--no matter how remote. Onboarding, conflict checking, kickoffs, and client briefings are all handled securely online and reviewers have the flexibility to commit to projects which suit the workers best. Communication, repeat engagements, mutual accountability, and flexible schedules are a few reasons why TCDI team members stay. Many have joined our staff full-time. We welcome them with open arms.

Dave Mattera says he fills his week working with the finest company and the most wonderful team of his career, volunteering at his church, overseeing a youth group, coaching softball, and spending time with his family. At work, he is best known for his grit, incredible enthusiasm, use (some may even call it overuse) of capital letters, and, of course… his trademark holiday countdowns!

His Work Mantra: "A Good Plan executed today beats a Great Plan next week" – Patton

Senior Technical Solutions Architect **Marshal Hagen** is heavily involved in Special Olympics in the Seattle area. In 2018, his sister-in-law's brother competed on the Alaska basketball team and Marshal wanted to utilize his love of photography to capture the moment. He applied and was accepted by Rod Mar (the official photographer for the Seattle Seahawks) to take photos for the week. During one game, Marshal sat next to a *Sports Illustrated* photographer at the edge of the court and picked up some professional tips.

Marshal's teenage daughter Audrey has intellectual disabilities and wanted to find a sport she could compete in. Bowling was the answer! Marshal and Audrey compete as a unified team and he can use his photography skills at their events. Over the years, he has taken more than 50,000 photos. Their team won Regionals and received a gold medal and took a silver medal at States. Marshal also coaches co-ed softball for the Special Olympics the team was awarded a silver medal. Being part of Special Olympics is rewarding for Marshal because he sees that most of the athletes would never have an opportunity to be part of a team, compete, and be taught by great coaches.

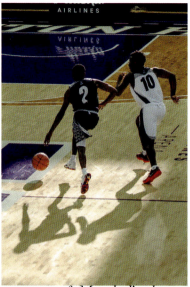

one of Marshal's photos

Marshal, Audrey, and their teammates Abbie and Bill.

17

UNCOMMON PHILANTHROPY

When the COVID pandemic hit in March 2020, Bill Johnson held an "All Hands" staff meeting to check on everybody. He wanted to know everyone's state of mind and how we were coping with suddenly living in a pod and without constant contact with anyone outside of our immediate family. He called on everyone individually to speak up. Some described their newly crowded and boisterous living situation with all parents and kids at home 24/7, while others felt secluded and lonely if they lived on their own. Bill genuinely wanted to know about everyone's mental health and how he could help. This kind of sincerity was unparalleled compared to most corporate owners. If a staff member had to take a leave of absence to care for a sick family member, they were provided a safety net and job security.

In a magnanimous act, Bill asked all of us to think about our favorite charity and gave us each $1000 to donate. A hundred employees donated $100,000 across dozens of nationwide charities. Soon after, we formed teams to concentrate on food, housing, education, and safety non-profits. Each group had $25,000 to donate and we were able to take off work to volunteer at soup kitchens, schools, and with first responders.

COVID forced non-essential businesses to close while independent shop owners struggled to stay afloat without any customers. Hair salons locked their doors. Libraries halted their Dewey decimal system. A tattooed-staff member at TCDI mentioned to Bill that his ink master might have to close his parlor for good. Bill floated the tattoo artist a $5000 loan to stay open until the man's customers could come back. A local barbershop faced the same situation and Bill gave the owner $500 as a "prepayment of haircuts." Bill helped because he could.

In the TCDI Cleveland office, our staff assisted with making 1200 meals at Dante Restaurant every Sunday to give out to folks who had lost their jobs. Our legal team was actively involved with chambers of commerce to work with local and state legislature to create laws that made it easier for companies to bring their staff back to work under the constraints of the pandemic. Businesses were afraid to re-open because they were afraid of getting sued by a customer who *might* get COVID. Companies that acted reasonably were protected through CDC and local health department recommendations.

Other staff found that the Winston/Salem Urban Housing Development and 1st Home Foundation needed donations. Local needy families were given mattresses, bedding, and linens so that they no longer had to sleep on the floor. Nearby law enforcement officers were given down payments for homes. We also helped build homes for underprivileged families.

To help a local lower-income elementary school, we donated money and supplies for a Back-To-School Drive. During Teacher Appreciation Week, we sent in breakfast for all of the teachers. During the holidays, we fulfilled entire Christmas wish lists with bikes, toys, and games for several families in need. Parents were given gift cards for groceries and gas. We helped because we could.

In 2010, after losing his beloved labradoodle at the age of seven to cancer, Bill started The Fred Foundation. The purpose of the foundation is to support organizations that seek to promote a more understanding and caring world in terms of human-human relationships and human-animal relationships. Since its inception, The Fred Foundation has given out more than $100,000.

We are strongly committed to participating in and positively impacting the community where we live and work. TCDI sponsors many community and charitable organizations such as Wyndham Championship, Pro-Am, The First Tee of the Triad, Kevin Harvick Foundation, Forge Greensboro, Friends for An Earlier Breast Cancer Test, Winterlark, Leukemia & Lymphoma Society, Operation Smile, Family Services of Piedmont, The Community Foundation, GTCC Golf Classic, Piedmont Triad Charitable Foundation, NCCJ of the Piedmont Triad, etc.

We are strongly committed to participating in and positively impacting the local community. Each year through the TCDI Cares program, we volunteer our time, talents, and resources towards supporting organizations that promote equality.

At TCDI, giving back to the community is an integral and meaningful part of the corporate culture. Our goal is to enrich the community and leave it better than it was. The generosity of our employees and the hospitality of local communities are humbling. We have the opportunity to make a difference in the lives of others and believe we must do so. We carefully choose causes and organizations that align with company goals and culture.

One stellar example of our workforce's generosity was shown by Dave York (below left, center) who

 climbed Mt. Kilimanjaro with his brothers to raise funds for the Lymphoma & Leukemia Society.

Closer to home, other TCDI employees ran a 10K to help support the cause.

It's hard to quantify the impact TCDI and Bill Johnson have on the Greensboro community. Most recently, we provided furniture and office equipment to assist a local tech startup as well as furnishing the new Greensboro Police Department building.

At TCDI, it is important to not only enhance and improve tomorrow's technology but also tomorrow's world.

On the East Coast, **Ray Arizmendi**, Discovery Engineer Client Data Services, is a freelance shutterbug in New York City. He co-authored *New York City Capital of the World* and supplied pictures of the cityscape, waterfalls, autumn colors, and aligning the moon with landmarks for the book. Follow him on Instagram @musicman33us. Below are some of Ray's photos.

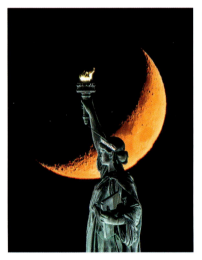  

18

THE DOOR IS ALWAYS OPEN

By Dave York, Managing Director, Litigation Services

As I write this, I have proudly worked at TCDI for nearly 16 years and have had the pleasure of knowing and working with Bill, Susan, and many TCDI colleagues as a client and law firm user for nearly 25 years. My wife, Melissa, started at TCDI a year before I did when she was my fiancé, and I was on the client side of the industry. When it was time for Melissa and me to move from fiancé status to husband and wife, Bill and Susan flew to the beaches of St. Thomas, along with a few of Melissa's TCDI co-workers, to not only be at our beach-front nuptials but to host a dinner for the family and friends that attended our destination wedding. Long before I joined TCDI as an employee, I was well aware of Bill and Susan's generosity as employers and how that culture of generosity and support pervaded TCDI's culture. That culture started with Bill and Susan and it has been present not just throughout my TCDI career, but throughout the growth of my family life.

During my years at TCDI, Melissa and I have had a wedding shower and two baby showers, with one of those babies now in high school. I could write an entire book about the different examples of Bill and Susan's giving and kindness, but there is one in particular that stands out for my family, and not just because it pops up in my photo memories each year.

After the birth of our youngest daughter and as our oldest approached elementary school, Melissa and I decided that it was time to sell our first family home and find the 'forever home' that would best fit our growing family needs. Selling and moving from a home at any point in life is stressful, but doing so with two small children and a busy work environment is a whole new level of stress. Very quickly, Melissa and I found ourselves with a pending sale and move date for our existing home, but no forever home to move to with our kids.

As our move date approached closer and closer, one day Susan stopped by my office and said that she had heard about Melissa and me selling our house. She offered us a solution to at least one of our problems – a place for our family to stay while we searched for our new home.

Susan and Bill owned a house about a mile from TCDI's Greensboro, N.C. offices, which was affectionately referred to as "The Dover Park Inn", because of its location on, you guessed it Dover Park Road, and because it was frequently used by TCDI for overnight guests so that they would not have to stay at a hotel. Over the years, The Dover Park Inn hosted TCDI employees, clients, friends, and even several PGA golfers during the annual Wyndham Championship. What Susan offered us was for The Dover Park Inn to become a temporary York home, so that we could eliminate the need to find short-term housing while we searched for our perfect house. We could store our furniture and worldly belongings in the garage and have free rein to settle into the house, eliminating at least one of the enormous stresses of the home search process.

And so, for several months, Melissa and I, along with our two young daughters and seventy-five-pound Labrador mix dog, called The Dover Park Inn home. We celebrated Christmas and birthdays there while our current home was being built. Our youngest daughter took her first steps in the kitchen at Dover Park, and those steps quickly turned into WWE-style jumps onto the living room couch. The girls played in the driveway and yard, and fell asleep on us while watching TV, and not once did the home feel like someone else's.

Every year, timeline photos and videos appear on my phone reminding me of those early days in our family's history, and the generosity of Bill and Susan. They have always had an open-door policy at the office and their homes have been no exception. They opened the door of The Dover Park Inn to my family, but this is not the first, nor is it the last time they have opened their home to their employees, friends, and colleagues. Bill and Susan's kindness in opening their homes to those that need it and the compassion that they show for their employees are at the core of the corporate culture at TCDI. I'm not only thankful for everything they have done for my family over the years, but I hope to emulate it and pay it forward in my own life and career. It starts with Bill and Susan, but it permeates throughout TCDI because we all know that the door is always open.

19

THE INDUSTRY IS NOTICING

TCDI and Bill Johnson have won multiple technology awards. As a small company, we never expected to run with the big players–the international firms who employ thousands of people. To be in the same conversation as Morgan Lewis and Epiq is a privilege.

Here are some of the awards that TCDI and Bill Johnson have won:

Lifetime Achievement Award at Legalweek Leaders in Tech Law Awards – April 2022

We are thrilled to announce that our CEO, Bill Johnson, was awarded the Lifetime Achievement award at the 2022 Legalweek Leaders in Tech Law Awards on March 10. In addition, TCDI's Military Spouse Managed Review (MSMR) program was shortlisted as a finalist in the Litigation & Discovery category. There is no one more deserving of the Lifetime Achievement award than Bill Johnson. Bill's vision and innovative culture show in our people, our software, our services, and our commitment to our clients. Bill's positive and consistent contributions to the legal technology community, the TCDI family, and the communities in which we live and work are significant and should be celebrated. Bill is a CEO whose legacy is about what he created around himself for others, not himself. The Lifetime Achievement Award recognizes Bill for his thoughtful leadership over the past three decades, his contributions to the legal technology community, and his ability to guide TCDI through industry transformation and expand service offerings to include large-scale litigation case management, e-discovery, digital forensics, and cybersecurity services, always with enthusiasm for his people and the clients we support. Following his win, ALM asked Bill about his work over the past 33+ years, his penchant for innovation, and more. Here's what Bill had to say.

Law.com: How has legal technology changed between when you entered the industry and now?

Bill Johnson: I actually don't think the core technologies—search, review, processing, and production—have changed substantially over the last 34 years. These technologies have remained fundamentally the same, other than the adaptation to the evolution of electronic data, social media, and the like.

What has changed, in a dramatic way, is the exponential growth of data volumes. In order to adjust to increases in scope, technology changes in hardware, as well as software architecture, have been critical to being able to "stay afloat" while keeping the technology responsive to business needs.

What does "innovation" in the law mean to you, and where do you think innovation is going from here?

It's tempting to think there is some paradigm-shifting technology under development that would fundamentally change how legal technology works, but the reality is that those things just do not happen, in any industry.

Instead, I look toward "useful" innovation where the technology does what technology does best—taking the human and providing tools to automate mundane tasks, improve the efficiency of tasks, and provide insight into large volumes of data. Ultimately, freeing the human mind to be creative and problem-solve.

What's one thing that you're most proud of over your time in legal technology?

What I am most proud of at TCDI is our ability to invest in and develop human capital to solve our clients' needs. Legal technology has always been, and will remain, about the people. It's an exercise in how to use technology to make the people more efficient, precise, and creative. We have been able to bring together the best minds in software development with the best minds in project management while remaining focused on finding the best way to capitalize on the knowledge and capabilities of the people in litigation support.

Best Places to Work – 2018

Triad's Best Places to Work by the *Triad Business Journal* in Greensboro/Winston-Salem

TCDI's culture is cooler and more collaborative than other companies. Imagine driving to work on Monday morning and looking forward to what the day has in store. Open the door and get greeted by four fluffy and friendly dogs. Stop by the indoor golf simulator and make a couple of jokes with the coworker who is trying to beat the top score. As mid-day approaches, the CEO calls an impromptu town hall meeting to discuss upcoming initiatives. He also grills a steak lunch for all employees. The TCDI office includes more than 700 feet of whiteboards, a large fun-room for meetings, several common spaces, comfortable seating for casual conversations, and outdoor patios. Flexible work schedules let employees attend school functions, volunteer, or create a better work/home balance. "Cheeseburger in Paradise" by Jimmy Buffett is TCDI's culture theme song.

Triad Business Journal Most Admired CEO - 2018

In this article, a TCDI employee described Bill at a trying time in his life. "His dedication and commitment to his employees and to the company is evident in everything that he does. Recently, he had to balance being with his immediate family during a severe illness or helping his TCDI family with a critical business development presentation that he was leading. He had a choice to be in DC with his family, or attend the presentation in Toronto, Canada. Amazingly, he did both. He flew with his team to Toronto and attended the meeting, and he did so with a heavy heart and his family on his mind. He made an amazing contribution to the presentation, captivating the room with each word that he said, and he left the meeting early to immediately fly back to DC to be with his family. After he left, several of the people attending the presentation came up to us to say how much it meant to them that Bill attended the meeting, especially in the middle of a family illness. The two of us from TCDI that also participated in the meeting were amazed at what we saw, and we were motivated to do an exceptional job for one main reason - we wanted to do it for Bill. It was inspiring to see someone push through personal pain to support his employees and his company."

Another employee added:

"In December of the first year I came to work for TCDI, Bill sent out a company-wide email where he mentioned every employee and what he appreciated about them, admired about them, and wanted to thank them for. Every employee. We were smaller then but not that much smaller! It was so obvious that it was from the heart and that he had taken a great deal of time thinking it through and making it personal to each person. I had been at TCDI for less than a month and he even had something to say about me. Bill is so quiet and unassuming but that should never be misinterpreted as his being out of touch with us, who we are, and what we do for this company. He knows all and he appreciates what every one of us contributes to TCDI."

And finally:

"Bill's 28% contribution to the company's ESOP in one of TCDI's least productive years speaks volumes. He has always been generous when times were good, but it is what he does in the harder times that shows his true character. He has always said that he feels personally responsible for 74 mortgages, and he shows it by always looking out for the financial well-being of his employees."

While talking about Bill's incredible reputation and philanthropic approach to the community that is not what makes him deserving of the 2018 Most Admired CEOs. His accomplishments are important but what sets him apart is knowing he truly cares about the success of every person that makes up TCDI. Bill Johnson is a CEO whose legacy is what he created around himself for others, not about himself.

Bill takes workplace culture seriously and invests time and resources to make sure he is constantly shaping a facility that benefits us employees, clients, and stakeholders. TCDI is a family and that starts with Bill. He has always taken care of employees the same way that most people take care of their own families. It takes less than

10 seconds to walk in the door and understand TCDI's culture. You could be instantly greeted by Bill's dogs or the dogs of other employees. He creates a relaxed work environment, which is great for a focused industry like ours. He believes in hiring top-quality, intelligent people, then giving them a small amount of direction and getting out of their way.

Bill creates and maintains a positive work environment through the many perks he has made commonplace at TCDI. Dog-friendly offices, an on-site gym with a full-time personal trainer, state-of-the-art facilities and resources, drinks and snacks provided by the company, flexible schedules, company-funded health insurance, company 401k contribution regardless of individual contribution, and unlimited time off for all full-time employees are a few of the examples of how Bill creates and maintains a positive work environment.

Transparency is important to Bill, and as such, he openly discusses high-level business plans with the entire company through regularly scheduled full-staff meetings. He sets the tone from the top, hires selectively, recognizes and rewards successes, and celebrates wins. He fosters open communication and dialogue throughout the organization regardless of title or tenure.

While it is easy to talk about a positive work environment when you are in it every day, it's even more impressive to hear it from those outside the organization.

This was shared with us by a TCDI client:

"As the person at my company responsible for managing a nearly two-decade-long business relationship with TCDI, I have had countless opportunities to work closely with Bill Johnson and his TCDI team. The reason our business relationship is nearing 20 years is simple: Bill treats his people with great dignity and respect. Bill's people who manage our relationship on a daily basis, not surprisingly, afford me and my team that same dignity and respect. A relationship predicated on dignity and respect is a winning relationship. A work environment predicated on dignity and respect is a winning, positive work environment. While I don't report to work every day to TCDI's offices, I could not be more certain that Bill Johnson has built and is maintaining a positive work environment that he and his employees are proud to call their own."

It's amazing to see Bill's thoughts and insights turn into vision. In one afternoon, he will compose a 10-page document outlining an entire system and process architecture, providing a roadmap for TCDI's next few years of software development. He communicates a vision, empowers his team to execute, and ensures they have the resources to be successful.

For the past 30 years, Bill has ensured that TCDI is leading the industry in litigation management innovation. He promotes a client-driven technology approach that allows us to meet the needs of existing clients. We stay a step ahead of industry trends, which in turn has allowed TCDI's CVSuite to thrive for over 30 years.

While the development of an innovative culture requires a strong foundation rooted in proven methodologies and theory, it cannot sustain without a top-down investment in the people who do the work. TCDI believes this is best accomplished through education, empowerment, and celebration. Every team member is trained through the basics of Lean Six Sigma and Agile philosophies. Bill has invested resources for over 5500 hours of training, resulting in one Master Black Belt, 13 Black Belts, 16 Green Belts, and 50 Yellow Belts. Perhaps more important than the training itself is the empowerment he gives his team to use it. Everyone at TCDI is encouraged to apply what they've learned through mapping processes, creating baseline measurements, and constantly asking why. To question the status quo is as commonplace as coming to work. Our workforce gains momentum with each stride and success. Bill seizes every opportunity to celebrate the wins through personal and public recognition.

Bill creates an innovative culture by hiring the best people for the job and creating an environment of mutual trust. He fully embraces their ideas,

Yetter Coleman - a litigation boutique specializing in high-stakes business and technology litigation around the country

"Yetter Coleman was introduced to TCDI's Military Spouse Managed Review Program in early 2019. Most recently, we have been pleased to partner with their review team in service to one of our clients in litigation. For this particular matter, TCDI compiled a team consisting of 100% military spouses and affiliates, including Project Manager and MSMR Program Leader, Jennifer Andres. Their accountability, communication, turnaround time, and daily reporting continue to impress us and help us push the ball forward for our clients – and we are excited to support members of our nation's armed services through remote work programs like TCDI's MSMR."

Stephani Brannan, Attorney, Yetter Coleman

thoughts, and opinions. For those of us who live near headquarters, an innovation campus was created with modern, open offices intended to promote increased collaboration and the flow of ideas between departments.

In the words of our TCDI family members:

"Bill creates and maintains a positive work environment by elevating "investing in employees" to an unmatched level. This goes far beyond mere words, training, or even monetary compensation."

"Bill creates an innovative culture by bringing in some of the best and brightest minds in the industry and fostering an environment of creative freedom."

10 Best Legal Tech Solution Providers to Watch in 2022 by *CIOCoverage* – Aug 2022

TCDI has been recognized by *CIOCoverage* magazine as one of the "10 Best Legal Tech Solution Providers to Watch in 2022."

Credit: *CIOCoverage*

"One of the leading tech solution providers, Technology Concepts & Design, Inc (TCDI), believes in transforming the tech landscape. Having been in the legal tech industry for more than 30 years, TCDI aims to provide its clients with the best legal services technology, with security, service, and innovation as its primary focus. With a platform that supports client-driven innovation and efficiency, the legal tech company initially started by creating customizable litigation technology solutions and web application development solutions to build and solidify the base of CVLynx, their proprietary Litigation Management, and eDiscovery platform. TCDI also believes in maintaining strong and transparent relationships with its clients and uses their resources to create practical and unique solutions for the challenges in the legal tech sphere.

TCDI identifies the challenges in the legal tech space and the trends that influence it. The minority-owned, US-based company keeps Lean Six Sigma principles at the forefront. It makes continuous efforts to invest in the best resources, transform processes and technology, and enhance its workforce to tackle difficulties legal teams may face in the discovery process. It provides solutions for issues such as social media data review, legal hold workflow, and collaborative team messaging data review.

The firm also aims to explore new and innovative methods to tackle text search handling, third-party cloud service integration, and data breach.

Clients often face difficulties navigating the legal tech sphere, such as individual data types and highly complex data relationships. TCDI, with its people, processes, and technology, identifies, and recognizes these challenges, organizes the unique data types, converts them into accessible data sets, and presents a simplified version to their users. The open-standards-based search solution, developed by TCDI, powers the CVLynx application providing precision full-text search that carries the range to control relational data structures on a grand scale to offer personalized features and services to clients."

"We believe working alongside our clients as partners foster strong relationships. Caring for colleagues and clients and 'doing the right thing' creates a culture of respect, safety, and accountability that is key to our success and, often, that of our clients," said Bill Johnson, CEO and Founder of TCDI.

The article went on to say, "The tech solution company provides solutions to data-organizational challenges and offers unique, top-notch legal and cybersecurity services. TCDI collaborates with clients to create a tailored and client-focused experience, one rarely found in the legal tech space. TCDI's team members offer their insight, knowledge, and expertise to keep the software up-to-date and ensure all the services run smoothly. Instead of incorporating temporary resources for quick and short-term fixes, TCDI's vision entails long-term planning that involves polished technology and customized methods to make the client experience more efficient, convenient, manageable, and flexible. For instance, CVLynx offers clients a way to create case profiles, transcripts, calendars, collect data, track and monitor processes, search, conduct operations, cull, classify and predict coding answers, and review documents to simplify the management of complex litigation.

Over the last 30 years, TCDI has been essential in making its clients' discovery process smoother. It has provided tools for legal teams in large-scale product liability cases in tobacco, firearms, pharmaceuticals, and multidistrict litigation cases in various healthcare spheres. TCDI operates on two principles to give and receive transparency from its clients: superior service and client-focused technology. Its resources are always at the disposal of its clients whenever they may require assistance."

TCDI's proprietary legal technology solution, CVLynx has been named to *The National Law Journal's* 2022 list of Legal Technology Trailblazers

This annual list, developed by the business arm of *The National Law Journal*, recognizes a handful of companies and products that are using innovation to help ease the daily tasks taken on by law firms and legal departments.

This distinction recognizes the power of CVLynx as a relational database designed for more than one purpose. Since 1988, CVLynx has supported the entire litigation cycle for some of the largest cases in U.S. history through core eDiscovery and Litigation Management functionality. The highly flexible structure of CVLynx has allowed it to adapt to other key client needs including core business functions, such as hours and invoice tracking, regulatory reviews, and submissions, as well as data breach reviews, analysis, and breach notification processes. "CVLynx is not an eDiscovery tool or a litigation case management tool or a data breach response tool, it is all three of those and more," said Caragh Landry TCDI's Chief Legal Process Officer.

Other selected recognition

Legalweek Leaders in Tech Law Litigation & Discovery finalist shortlist 2022: Military Spouse Managed Review (MSMR) program

ILTA Distinguished Peer Awards Lifetime Achievement Award Runner-Up 2021: Allan Crawford

CIO Review, 10 Most Promising Legal Technology Companies 2020

Insight Success, The 10 Most Innovative E-Discovery Disruptors of 2018

CIO Review, 20 Most Promising Legal Technology Solution Providers 2018

Senior Programmer/Analyst **Eric Anderson** loves playing games! Cards, dice, and board games, that is. He and **Paul Frediani** in TechOps have been friends for decades and they formed a love of games back in high school. Along with other buddies, they have played almost every week since. Eric says it's satisfying breaking open a new game, grasping the rules, and forming strategies with teammates. His favorite games are Return to Dark Tower, King of Tokyo, Dragon Dice, and Clank Legacy. He, Paul, and their fellow gamers play three-to-four games per night and the only prize they go home with is bragging rights. More often, they simply play for fun and don't even remember who won.

Around the table: Paddy Hanner, Rob Genadio, Paul Frediani, Christine Anderson, Eric Anderson, Brian Hanner

20

WHY **TCDI**

By Andy Cosgrove, Chief Strategy Officer

As data piles up and litigation deadlines approach, it sure is easy to feel like eDiscovery is just Lucy and Ethel's historic battle to keep up with the chocolate factory assembly line in the classic *I Love Lucy* episode. Go until you can't keep up – break down – stop the line – clean up – start again.

After nearly a decade as in-house eDiscovery counsel, there have been days where I have seen the best systems and teams pressed to their limits – people without sleep, data without end, and technology stretched past its capabilities. But we stay, we pick up the pieces, we re-assess, and we move forward. As time passes from matter to matter, we sleep with document counts, workflow charts, and coding panels floating in our dreams.

Are we really Lucy and Ethel though? Are we standing and waiting for the next batch of chocolates/documents to come down the line waiting to be wrapped? Is it all just a 'losing game?'

Of course not. But how do we set the conditions necessary for sustainable success? How do we scale, flex, and deliver excellence and efficiency under the data onslaught?

It all comes back to a matter of unflinching alignment to a single goal – solve the problem.

Solving it today is better than solving it tomorrow. But the real value is not derived from solving today's problems today. Instead, it is from solving tomorrow's problems today.

How does an eDiscovery team ever put itself in a position to do that?

In the end, it is based on the recognition that looking forward to solve tomorrow's problems requires trust. Trust that the 'chocolates' will keep moving down the line today neatly wrapped into their boxes for delivery. With that trust comes a recognition that the team must be built so that forward-looking problem-solving is supported as a critical continuous operation.

That vision must be matched with creativity and given freedom so that our collective eDiscovery pasts cannot establish an incontrovertible dogma that limits the options for our best future.

Every bit of added value in legal risk mitigation and cost-savings starts there – solve the problem.

In this space, few truly do. Most, at best, throw bodies and servers at the problem – rarely looking forward and never looking back. Without a true foundation of vision, innovation, and sustainability, it is far too easy to fall back on brute force as the only tool that could possibly serve.

After working with TCDI for three decades, I have seen TCDI's relentless optimism that their clients' intractable problems will be interesting to solve.

Solve tomorrow's problem today. It is why clients come. It is why clients stay.

It is why I am excited to join TCDI's amazing team and fight the 'winning game' for TCDI's clients.

21

Getting Back on the Horse

Challenges happen. It's a way of life. At TCDI, sometimes we must handle difficult situations and people. The legal world has extremely demanding and changing timelines. To survive and succeed, companies need to move at a rocket's pace.

Larger clients didn't initially buy into TCDI, because we have less than 200 employees–half that in 2017. They wanted strength in numbers. The potential clients didn't believe that a small company like TCDI could out-perform, let alone even handle, massive terabytes of secure data. We had to prove ourselves time and again.

Sometimes it took two years for our sales staff to secure a client. A client's existing vendor may have screwed something up and the client was forced to make a change. As a result, we had to fix the problems that the client's former vendor created. We could initially be in the hole in terms of sales coming in but would be in a better position for the future with the new client. That was the risk we were willing to take.

With that, losing clients is always a threat in the tech industry. We have lost our fair share with a punch to the gut but kept moving forward.

Once established, client projects may have a demanding and heavy trial schedule. Legal operations need to track exhibits and documents. Within the industry, it's a struggle to keep up with all of the federal and state policies. We learned the most at those times trying to come up with creative solutions for our clients' needs. We ask, "How can we automate it for you?" with continuous process improvements and provide solutions.

Time is our nemesis. Our clients want results yesterday. They request 50 terabytes of data to be secured and wish it could be done overnight. We have to work around those inherent challenges. Having a 24/7/365 data center alleviates some of those obstacles. As a benefit, our networking team can make immediate changes without waiting on a third-party vendor. We can do it faster and understand the depth of our clients' systems. A third-party vendor might not have that expertise.

Conversely, having an in-house data center has its challenges. Instead of hosting data on the cloud, we have servers and generators that run 24/7/365—and, consequently, can break down. Our eight data center techs must maintain all of the networks and storage. Additionally, they must stay vigilant when it comes to the weather. Gale-force winds, rain, and even hurricanes could wipe everything out in an instant. We remained primed as much as possible.

Our employees must rise above the weather at times. Occasionally, we have snowstorms in North Carolina. In a state where schools are closed when an inch of snow falls, we have to prepare for the worst. One winter, a rare blizzard hit the same day as a major data migration. Power and driving on roads were sketchy at best. One staff member who lived within a few miles of the office brought a generator and extension cords to keep the migration running while he stayed overnight at TCDI. Another team member attempted to face the snow but wound up stranded on the side of the road. A Good Samaritan in a Humvee towed him to the office. The migration was complete, despite the dump of snow from Mother Nature. As a thank you for their dedicated efforts, Bill catered lunch for our migration team. On other occasions, we chased flying documents in a windstorm and slept on couches at the office because computers failed.

In the tech field at other corporations, employee turnover is at a rapid pace. Top-notch developers might stay at a company for a year and move to the next one, taking all

Goldberg Segalla - a national civil litigation firm with more than 20 offices in 10 states spanning major metro markets across the U.S.

"We've finished our doc review and are prepping for production on this one. I wanted to shoot you a thank you note – you and your team were absolutely great. This has been the smoothest outside doc review project I've done (out of 30+ over my career so far), even considering that we shifted platforms halfway through (and thank you for that, that was huge). Rebecca and Ned went above and beyond every step of the way."

Jeffrey Cunningham, Esq., Goldberg Segalla

of their knowledge with them. We have to keep our core of people interested and engaged as technology changes. If products don't stay abreast with the industry, we might face 15-year-old technology. We need to be challenged and technology needs to be fresh. We constantly try to stay ahead of the curve.

But not all circumstances had a happy ending for us.

In 2017, our confidence took a hit. We had the opportunity to shine with a massive multi-million-dollar project. We won the bid to process, host, and support terabytes of data for discovery in 2015. We built a highly custom and highly secure system that could handle a unique, automated workflow that was unprecedented in the industry. Two years into the endeavor and 62,000,000 documents later, a larger competitor that was collaborating with our clients found a way to terminate our contracts, move the data away from us, and cast a shadow on us. We got stomped on by a giant with more funding and backing. Lesson learned: we saw an opportunity for innovation but over-engineered the solution based on the competition's requirements (not our clients' requirements) and made it too complex. We listened to the wrong source of information and paid the price. Even though we still made a profit, we lost future potential relationships. We learned from our mistakes and about ourselves (overly-complex processes, corporate boundaries, and industry relationships) and moved on.

That unexpected hit only made us better.

Receptionist **Christine Kluge** owns Christine's Gentle Touch Dog Grooming. She has always loved dogs. When she was a little girl, her constant canine companion was a Labrador Retriever named Butch. Butch never left her side and only let Christine's parents near her baby carriage.

Jump ahead to her adult life… After getting laid off from AMP Inc. after 20 years of service, Christine embarked (pardon the pun!) on a new career. In 2001, armed with her severance pay from AMP, she took a Canadian online dog grooming class. After six months of intense training, she graduated with AKC and CKC certifications and bought equipment to start her new dog grooming business.

Christine's clients (including Bill) have mentioned that her business is "A Spa for Dogs." Soothing classical music relaxes the dogs as they are pampered. Some of the pups have fallen asleep on the table. All pooches are groomed by appointment only so each dog receives the special attention he or she deserves. This appointment-only approach cuts down on barking and stress for her precious four-legged friends.

Some of Christine's clients: Chiko (17 years young), Frank, and Luna

Vice President of Cybersecurity **Eric Vanderburg** fondly remembers the times he spent in arcades as a child. The traditional arcade is mostly gone, but Eric keeps the spirit alive by building and refurbishing arcade machines. He currently has eight arcade machines, a pinball machine, and a pachinko machine in his basement arcade and is in the process of refurbishing two more 1970s pinball machines. Eric has built many others for friends and charities, such as Laura's Home Women's Crisis Center and Grace Baptist Church.

22

WELCOME HOME

By Joe Anguilano, Managing Director, Cybersecurity

My eDiscovery and professional career began 16 years ago when General Counsel Tim Opsitnick hired me right out of college as a Business Analyst for the eDiscovery and forensics company he founded called, JURINNOV. Right from the start, it was clear how much he cared for the people working for him, myself included. For the next 10 years, I had the pleasure of working for Tim in various roles and capacities. I learned a great deal, had fun and challenging work, and loved my job.

Then, in 2016, JURINNOV was acquired by TCDI. I was lucky enough to play a key role in the work leading up to the acquisition with documentation gathering and other due diligence. More great experience. But still, I had some apprehension about now having a new employer. JURINNOV was my first and only job since graduating college 10 years prior.

I found myself thinking, "How could things get better than they are now?" Tim was an amazing boss and employer. This is someone who, as a bonus, sent my family to Orlando on vacation, helped me get my MBA, got me custom-tailored suits as a graduation present, and gave me an unlimited amount of vacation time to use when my newborn son was sick and in the NICU (doing great today). We'd been to Cleveland Browns games, Guardians games, and he was at my wedding.

Though I was naturally apprehensive about a potential acquisition, one thing that put my mind at ease was this: One of the motives for selling to TCDI was that TCDI provided additional opportunities for JURINNOV employees to improve their lives both professionally and personally.

But the reason I'm here is not to focus so much on JURINNOV but rather on my experience during the acquisition and where some of the other JURINNOV team members and I are today.

One of my first experiences after the sale back in 2016 was a trip down to TCDI HQ in Greensboro to meet senior managers and the rest of the team. I had the warmest welcome and was met by some of the kindest, most considerate people you could hope for. The first two integration projects we worked on included data migration from our collocation facility to Greensboro and invoicing. These migration projects were a great way to start working with the new team.

Other than those two items, it was business as usual from a client services perspective. We operated for a while as JURINNOV – A TCDI Company, until fully moving to simply TCDI.

Now, six years after joining TCDI, can I say that it has improved my life both professionally and personally? 100% yes. But it's not just me. My colleagues did as well. An Administrative Assistant joined JURINNOV in 2015 and is now a Marketing Specialist with TCDI. A System Administrator joined JURINNOV in 2015 and is now TCDI's Senior Cybersecurity and Forensics Engineer. A Senior Network Engineer joined JURINNOV in 2006 and is now TCDI's VP of Cybersecurity.

I feel lucky because TCDI not only provided me with opportunities for advancement but is also an employer who cares deeply about its people. I see employee morale and well-being factored into every decision. For example, during a management meeting in Greensboro, Bill Johnson told everyone in the room that they could spend up to $1000 on their respective teams. For the Cyber team, we picked a Friday and met for lunch at the Corner Alley which is one of those fancy bowling alley/restaurant combos. Eight of us got together and had a blast. Then we planned a BBQ in July.

Since graduating college 16 years ago, I've worked for two companies JURINNOV and TCDI. And two bosses: Tim and Bill. I sometimes find myself reflecting back on that and ask myself: How did I get so lucky? Even with the limited number of employers I've had, I know a good thing when I see one.

Today, I have the privilege of working with an amazing team and have access to so many resources. TCDI invests in having someone of this caliber on staff.

That's my story.

Marketing Specialist **Katie Niemi** and her husband, Mitch, rebuilt a 25' Hunter sailboat. The ship was in bad condition after it took on water during the winter months. The previous owner lived in another state and had no way to remedy the situation. After several discussions, Katie and Mitch agreed to take ownership of the boat and get her back in the water. After a summer's worth of work, they replaced large portions of the inside hull, fixed the parts of the decking, and replaced broken equipment. They now cruise around Lake Erie and have also taught some of their coworkers about sailing. Who knows, one day Katie might transition from the work-from-home to the work-from-sea model.

QA Director **Sandy Middlebrook** has always loved animals. She and her husband Jay have owned CenterStream Farm since 2005. The 43-acre farm in North Carolina has hosted various animals over the years including horses, miniature horses, llamas, goats, cats, guinea hens, and a rooster. At one point, they had over 80 animals on the farm. Nearby wildlife likes to check out the property so Sandy and Jay's goats are protected by two Great Pyrenees livestock guard dogs named Coco and Ghost. The couple also owns rare Sussex Spaniels named Trumpette and Bailey who frequently let her dress them up for holiday pictures.

Sandy is also an avid photographer. She and her grandson Nathan have a weekly photoshoot with him in front of and behind the camera. Sandy has taken many corporate photos at TCDI and often posts pictures of mornings on her farm on employee message boards. Her website is www.csfphotography.net.

Check out some of Sandy's photos.

UIUX Designer **Lisa Mays** has a large brood.

"Don't tell Mom!" her son Raymond said recently, then laughter erupted around her holiday table. Many years after the fact, Lisa was now privy to stories from her 10 kids that make her mouth fly open.

Yes, 10 kids. But don't be alarmed. She didn't give birth to 10 kids, but she and her husband Ed *raised* 10 kids. They have a blended family. When they got married many years ago, Lisa had two kiddos (one biological and one adopted) from her first marriage and Ed had four kids (two biological and two adopted) from his first marriage. Together, they created a *Brady Bunch* family of six kids. However, they both felt their family was not complete, so they added four foster kids to the mix.

In total, Lisa has one biological child, two stepchildren, three adopted children, and four foster children. She remembered a specific period when every single kiddo in the house was a teenager (ages 13-19). In addition, they added one exchange student from Spain every year for four years to give their kids the opportunity to learn about another culture (and vice versa). Those were crazy and creative days.

Managing 10 teenagers in fluctuating hormonal stages was not for the faint of heart. Raising eight boys and two girls required patience and creativity. One of Lisa and Ed's most creative ideas was born out of desperation. Exhaustion overcame them trying to figure out which kid left their towel on the bathroom floor, dishes on the table, or stole their sibling's hangers. Lisa and Ed developed a color system. Every kid picked their favorite color and the family went on a mission to find EVERYTHING (towels, sheets, plates, cups, utensils, and hangers) in those colors. Jessica chose purple, Steve picked black, Troy was yellow, Chardae liked blue, Nathan chose red, Raymond picked green, and so it went. From that point on, there was no question about whose towel was on the floor or who left a cup on the table. The kids quickly learned responsibility and accountability. More importantly, peace existed in the Mays household.

Today, Lisa and Ed are empty nesters. Their child-rearing is done. But when the kids come home for the holidays, they love reminiscing about the crazy adventures they had, and antics they pulled that typically ended with, "Don't tell Mom!"

Some of Lisa's large family. Top row: Raymond, Eddie III, Jordan (Chardae's husband), Eddie IV (grandson), Jhon (Jessica's fiancé), Bottom Row: Kimarah (granddaughter), Chardae, Jessica

Top: Rachel (exchange student from Spain) Middle: Raymond, Rachel, Jessica, Steve Bottom: Jessica and Rachel

Chardae (Lisa's adopted daughter) with the Clintons. She was selected as the "poster child" for National Adoption Day in 1997 and was invited to the White House.

23

THANK YOU FOR YOUR SERVICE

Not only on Veteran's Day, but every day, we like to honor our team members who are former and active service members. We resolve to thank them *throughout the year* for the vital service they have performed for our country. Our daily lives are better because of them. They set an example for all of us. We are thankful daily for the talent and leadership that these folks provide and the unique perspectives they offer TCDI. They have taken risks and made sacrifices that many of us cannot fathom. Our freedom isn't guaranteed. It is protected by these men and women who work among us and secure our way of life. Our country's greatness is built on the foundation of their courage. We are proud to call them our coworkers, our friends, and our family. We can never thank them enough. We owe them our freedom. We owe them everything.

Top row: Steve Wujek, Chris Attucks, Sterling Tysinger, Dave Scallion
Bottom row: Mark Zurovec, Anthony Klier, Geoff McPherson, Frank Bounds

Frank Bounds, who has been with TCDI since 2010, went into the Navy because they had a great reputation for technical training. His biggest takeaway from being on submarines was teamwork; everyone was valued as part of the team. Their lives depended on each other and each was expected to do their part. He was stationed on the *USS Sam Rayburn* (SSBN-635) based out of New London, CT, and stationed in Scotland. Frank retired as a Petty Officer 2nd Class and found employment with General Electric to run a highly specialized data center. In his spare time, he plays mandolin, banjo, and Dobro guitar in local bluegrass and Americana bands.

The submarine below was Frank's home-away-from-home for five years.

*Left: Technical Writer **Anays Quintero** (Army) earned her stripes in 2014.*
Right: With her husband in medic training in 2011.

Colleen Sweeney has been a licensed attorney for 25 years. Not only is she a military spouse, but she is also a Veteran having served eight years in the United States Army Judge Advocate General's Corps (JAG). Her husband served 20 years on active duty as an Apache helicopter pilot. They met while they were both stationed in Germany. They have moved 10 times with the Army and are the proud parents of four children. As they say in the Army, "Home is where the Army sends us." She has had the privilege of calling many places "home" including Ft. Bragg, NC, Ft. Leavenworth, KS, and Ft. Hood, TX. All of that moving created challenges for her professionally but she believes the trait of adaptability, which every military spouse possesses, has served her well in reinventing herself with each move. Simultaneously wearing the titles of "Veteran" and "Military Spouse" are the greatest honors of her life.

Above Left: Colleen and her parents
Above Right: Colleen's dad Col. (Ret.) Henry F. Coyne

Above: Colleen and her family at her husband's homecoming.
Below: Colleen's husband

Colleen and members of her unit

Not only do we honor our staff, but we honor the families of our staff who have served.

Left: **Ann Newton's** *dad in the Army in Germany. Right: At her husband's Air Force retirement ceremony.*

Garrick Muench's *dad*

Lisa Autry's *sons and future daughter in law*

Todd Kirmayer's *dad in the Air Force*

Left: **Milton Hooper** *stationed at Thule Air Base in Greenland. Right: Milton and his dad in the Air Force.*

94

Left: **Sandy Middlebrook's** *husband Jay on a submarine in Vietnam. Right: Sandy's son Chris in Iraq.*

Mary Walsh's *brother Jon was stationed on the* USS Maine *(SSBN-741) in the Navy.*

24

REMEMBERING OUR FRIENDS

On November 3rd, we celebrate Allan Crawford Day at TCDI. In memory of our dear friend, colleague, and mentor, we go outdoors, have a little fun, or donate to a cause that is close to them or their family. Allan lived his life to "do some good and have fun." Though he wasn't on earth long enough, he left an indelible mark on everyone he met. Allan had a sense of humor that lit up any room and kept those around him on their toes never knowing what crazy antics could occur next. We unexpectedly lost Allan on November 3, 2021, and we honor his life, work, and legacy.

Allan was a solutions-oriented, litigation technology professional with more than 25 years of industry experience having managed investigations and electronic discovery for law firms, corporations, and government agencies. He was a Senior Solutions Architect at TCDI.

At TCDI, migrations were one of Allan's fortes and he was working on an article detailing how to successfully move eDiscovery work from one provider to another, even during what feels like the worst time possible. Sadly, Allan passed away before he completed the article, but his TCDI family helped him finalize what he had started.

We honored Allan's memory by participating in the 2022 Midlands Heart Walk in Columbia, SC, on March 26, 2022, and raised $10,000 for the American Heart Association. He is survived by his wife Cheryl and children Zachary, Jacob, Kenneth (Kelsey), Julia, Hailey, and Chase.

Polly Hatcher was hard working with high attention to detail and great at her job. She was always smiling, with the kind of grin that would light up her entire face and you couldn't help but smile back. She was kind-hearted, strong with a fighting spirit, and always a positive and uplifting person to be around.

On March 26, 2007, Polly left us at the young age of 41 from complications of leukemia. A native of Savannah, she was a graduate of the University of Georgia and the Georgia State University School of Law. She was admitted to the Georgia Bar in 1990. Following employment in the Office of Domestic Legal Services of the Fulton County Superior Court, she was associated with King & Spalding, the Atlanta law firm, from 1998 to 2004. At the time of her death, she was vice president for Discovery Services at TCDI.

We deeply miss you Allan and Polly.

25

BEING GENUINE IS SECOND NATURE

At TCDI, we aspire for excellence and that starts with being genuine. As sincere people, we are usually more effective at motivating because we inspire trust and admiration through our actions, not just words. We walk the talk and don't think twice about it.

We don't try to make everyone like us. We are comfortable with ourselves as a company and understand that potential clients will hire us, and some won't. But we're okay with that. It's not that we lack empathy, but instead, we won't let someone else's feelings get in our way of doing the right thing. We are at ease with making unpopular decisions if that's what needs to be done.

To be an approachable and interesting company, we strive to be open-minded. Our clients are appreciative of having conversations with us where we listen to them without forming preconceived opinions. We want to understand their needs and goals. As a result, we have access to new ideas and better ways to attain our clients' goals.

We have forged our own path. As a company, we know who we are and don't pretend to be anything else. With our progressive round organization chart, we have principles and values. We are not swayed by the fact that someone else might not like it.

Our philanthropy demonstrates our generosity. We are giving with our time, our resources, what we know, and who we know. We want others to do as well as we do. Because we can.

We treat everyone with respect. Whether interacting with a founding partner in a law firm or the server who takes our dinner order, we always maintain politeness and respect. We do this because we believe we are no better than anyone else and everyone deserves the same treatment.

Our happiness comes from within. We are genuinely happy to come to work and be around each other. Simple pleasures such as friends, family, and a sense of purpose make our lives rich. That's why many of us have worked at TCDI for over a decade.

Our clients trust us. We mean what we say and deliver what was promised. Bill Johnson trusts us to do our job and rewarded us with an ESOP company.

We are self-confident in our words and actions. If a client criticizes an idea, we don't take offense to it. We take the hit and keep moving forward. There is no need for us to waste time or energy feeling insulted. We try to objectively evaluate negative and constructive feedback, accept what works, put it into practice, and leave the rest of it behind without hard feelings.

We want real conversations with our clients. We commit to a discussion and focus attention on our client. In effect, we create connections with our clients and find depth in short conversations. Our genuine interest in our clients makes it easy for us to ask quality questions and deliver their needs.

At TCDI, we believe we are firmly grounded in reality because we know who we are.

Programmer **Ty D'Angelo** plays premier-league soccer for the Celtic Cowboys in Austin, Texas. Even though Ty is right-handed, he plays left forward because his left foot is stronger. In 2022, he claimed the most goals on the team by a single player with seven goals. Ty enjoys playing soccer on the weekends to get some physical activity into his routine as well as socializing with his teammates and learning to be more competitive.

26

MENU OF SERVICES

CVUnity

CVUnity layers dashboards across multiple systems and datapoints throughout the litigation lifecycle in one platform while reducing legal spending. This is a product for corporate and litigious clients, specifically C-Suite executives, legal operations, associate general counsel, and outside counsel who need to take charge of their cases and litigation data, workflows, and resources to maintain cost control in litigation. CVUnity is a modular architecture application that links data from different business areas directly to CV hosted databases and maintains all data in one location.

CVUnity litigation management gives the ability to view report statistics, maintain a data calendar and tasks, manage matters, etc. Additionally, CVUnity has capabilities for expansion, integration, customization, and scaling the litigation workflow management.

Users save time, effort, and costs by utilizing CVUnity with standardized processes. It brings parties together to communicate common goals set by corporate counsel, reduces repeated or inconsistent work product from being created, and increases efficiencies and throughputs within litigation teams.

CVLynx Technology Assisted Review/Continuous Active Learning

Technology Assisted Review (TAR) and Continuous Active Learning (CAL) are advanced computer-assisted review technologies offered by CVLynx that leverage machine learning models, binary classification algorithms, and quantitative analysis of full-text content for each document. Powerful and intuitive like the big cat it's named for, CVLynx is a scalable eDiscovery platform that enhances document review productivity and efficiency for complex litigation. It quickly organizes data sets with email threading, near-duplicate identification, and conceptual clustering. CVLynx is compatible with most document reviews and accessible at almost any time. It leverages analytics to offer document review workflow strategies. These include review and Quality Control (QC) coding suggestions, tagging predictions, ranking, and prioritizing documents into targeted review and QC subsets, automated coding, and more. The CVLynx TAR/CAL workflow premise is to have human reviewers work as they normally would in any other document review while a predictive model for each classifier is automatically created and routinely enhanced in the background. The tagging decisions made by the human reviewers serve as inputs or training examples to enhance the classifiers in the predictive model. In turn, the predictive model provides several actionable outputs based on the desired strategies of the review administrator for each classifier.

Cybersecurity Services

Our innovative approach to cybersecurity services is a winning combination of innovative technology and hands-on client support. Our organization's dedicated staff of trusted advisors, security experts, and industry thought leaders rely on our proprietary cybersecurity assessment application to create custom-tailored solutions to meet our client's unique needs.

Litigation Management

Let TCDI help tame the chaos. Organizing massive amounts of information associated with legal matters can be challenging. In the past 30 years, TCDI has helped clients manage thousands of matters through litigation management platforms, working hand-in-hand to collaborate on their day-to-day demands.

Since TCDI is obsessed with the client experience, we're constantly looking for ways to help our clients be more efficient. That means designing and building a platform that is easy to use and customizable, with tools for case automation, evidence tracking, witness materials, pleadings, and all case materials ranging from key documents to deposition testimony. While many legal matters don't require a litigation management system as robust as CVLynx, it's designed to scale up and down to fit our customer's needs. TCDI excels at designing analytic systems so our clients never outgrow them.

When it comes to litigation management, experience is priceless. For three decades, we've quietly nurtured deep, lasting relationships with corporations involved in large, complex matters, in sectors that include Pharmaceuticals, Tobacco, Insurance, Manufacturing, Healthcare, and Banking. We developed and managed the litigation management system used by all parties in the World Trade Center Disaster Site Litigation.

Nextra Solutions - the information management and advisory service of the Nexsen Pruet law firm

"Having worked closely with TCDI in service to large corporate clients for many years, I was eager to continue that relationship when I joined Nextra Solutions as its Director. When presented with the opportunity to support TCDI's Military Spouse Managed Review program, we enlisted their team to assist us in handling an internal investigation with a quick turnaround. Our experience with TCDI has been nothing but positive and we will continue to find ways to work together in support of our country's military families."

Angela O'Neal, Director, Nextra Solutions

eDiscovery Services

Across businesses of all sizes and sectors, legal teams are under pressure to reduce the cost and complexity of the discovery process. That's where we come in. Our proven eDiscovery software and services, extensive industry experience, and superior client support enhance efficiency. Our services also deliver higher-quality results while decreasing costs. Our work further supports our clients by allowing them to have entire teams freed up to perform more important tasks, such as developing better case strategies.

TCDI provides electronic discovery services spanning the Electronic Discovery Reference Model (EDRM), including Collection, Processing, Analysis, Review, and Production. Our powerful, sophisticated technology helps clients reduce data volumes while enabling more efficient and accurate human review.

We provide a client-focused approach to managing matters, collaboration, increasing efficiencies, and reducing cost, all unified in one platform that provides predictability. Even though we're partial to our proprietary-hosted review technology, we fully support other major hosted review platforms as well, including Relativity and Brainspace, because we know that providing the best client service means working the way our clients want to work.

Penetration Testing

TCDI's penetration testing services discover internal and external security vulnerabilities through a simulated cyber-attack. Performed by our highly skilled experts using sophisticated testing tools and technology, this test provides unparalleled insight into security vulnerabilities that could be exploited and result in a data breach.

Following the completion of the penetration test, a TCDI team will provide a written report that summarizes the exploitation attempts, the vulnerabilities identified, their threat level, and the proper remediation steps to be implemented. At TCDI, we are proud to empower our clients to optimize their security through expert strategic advice and innovative technology solutions.

Cybersecurity Risk Management Program

The management of technology has seen both subtle and obvious changes over the last few years. This, coupled with an increasingly complex regulatory environment, heightened consumer concern over data privacy. A significantly higher impact of cybersecurity threats has led to a perfect storm of cybersecurity challenges that keep business leaders up at night.

Some have addressed these challenges by hiring Chief Security Officers (CSOs) and other specialized cybersecurity professionals, but this isn't always the right fit for an organization. These professionals are in high demand and command high salaries.

Cybersecurity is far too broad for a single individual to manage effectively. The skills and knowledge necessary to keep up-to-date with the latest threats while simultaneously minimizing risk is a mammoth undertaking in and of itself. As a result, teams are often required, which goes back to the last point: the demand far exceeds the available supply.

CV Case Management

Many corporations still manage all case-related information and documents in different systems and spreadsheets. This partitioned set of tools doesn't give the full picture. These tools must allow for connections to be seen among all these large, multi-districted matters. Many times, multiple outside counsel, different districts, and strategies aren't cohesive. When one district's strategy is "working", that strategy isn't shared or seen among other counsel. The lack of cohesion in the entire process can cost the corporation millions of dollars in losses, whereas better collaboration could bring everyone together.

Since CVLynx can store all case-related documents in a single repository, this allows a blend of eDiscovery data, exhibits, transcripts, profiles for plaintiff counsel, and judges all organized by jurisdiction to make real-time decisions. Having access to this information immediately gives insight into the jurisdictional outcomes of cases to impact overall strategy as the Multi-District Litigation (MDL) or Class Action grows. It also allows all outside counsel to have access to information so they can work together on their overall strategies. All data is organized and can be accessed with simple searches or drilled down in charts to get to what is needed quickly.

Cybersecurity Assessment

Evaluate a company's security and privacy against stringent globally recognized standards and best practices. The cybersecurity assessment can be used to validate adherence to relevant standards or as an easy-to-understand, prioritized road map for enhancing privacy and security.

Specifically, the cybersecurity assessment focuses on the following topics:

Breach Notification

Data Governance/Classification/ Handling

Email Security

Employee Training

Information Security

Inventory and Asset Management

Mobile Devices

Physical/Facility Security

Policies and Plans

Regulatory Compliance and Audit

Risk Management

Network Security

Software Development

Vendor Management

Backup and Recovery

Resiliency, Business Continuity, and Disaster Recovery

Encryption

Authentication and Access Controls

Logging, Auditing, and Monitoring

Vulnerability Management

Malware Protection

Patch Management

Endpoint Protection

Wireless Security

CyberPulse 365 and Secure Owl

CyberPulse 365 offers an affordable and accessible solution to protect critical data and their businesses' reputation.

TCDI's managed security service includes the following capabilities:

• Cybersecurity Monitoring and SIEM Platform: An integrated Security Information and Event Management (SIEM) platform provides robust cybersecurity monitoring and alerts. The SecureOwl appliance collects logs from devices on your network including servers, workstations, switches, routers, firewalls, and storage devices. It encrypts and sends the information to TCDI for analysis and archival.

• Threat Detection and Automated Remediation: Events are analyzed in real-time, and TCDI correlates information from various devices to gain a holistic understanding of the threat. Besides manual notifications, certain events can trigger automatic workflows to mitigate the threat such as disabling an account or quarantining a device.

• Malware Protection: Advanced malware protection combines endpoint protection, centralized monitoring, rapid virus definition deployment, and access to optional incident response and malware sandboxing services to provide a powerful defense against an attack.

• Vulnerability Management: CyberPulse 365 identifies issues with incorrect configurations, system changes, or software bugs so they can be corrected before they are exploited by hackers or malicious insiders. TCDI will scan client networks monthly and deliver a list of vulnerabilities and prioritize remediation actions.

• Data Loss Prevention (DLP) policies are enforced across devices to control how data is used, stored, and transmitted. Some actions may trigger an alert while others are automatically prevented, thus stopping data from traversing to unauthorized cloud services, external devices, or unknown email recipients.

Employee Data Theft Investigations

While employee data theft can happen at any time, it occurs most frequently before, or immediately after, an individual's termination, or resignation from an organization. Motives for data theft include setting up a competing business, using the information at a new job, a sense of ownership of what was created, and revenge against the employer.

The most commonly stolen intellectual property and trade secrets include:

- Customer information

- Financial records

- Software code

- Email lists

- Strategic plans

Virtual CISO Service: Chief Security and Privacy Officer on Demand

Cyber risk is a complex business issue, not simply an IT issue. Small and mid-sized businesses with immature cyber risk management programs tend to sweep cyber risk under the proverbial rug, hoping somebody is taking care of it. Many trust their managed services provider or in-house IT staff to handle the technical side of security. Unfortunately, managing cybersecurity requires specialized skills and experience outside of a typical IT employee's scope of work.

Bringing on a full-time CISO is expensive and recruitment is highly competitive in today's job market due to the nation's huge cybersecurity

Carswell Distributing - Bobcat sales, service, parts, and attachment needs

"Carswell contracted with TCDI to undergo a cybersecurity risk assessment to evaluate their overall IT risk and create a roadmap to address their vulnerabilities. Impact: Months after the cybersecurity risk assessment was complete, COVID-19 struck the United States drastically changing business operations as employers across the country suddenly required their employees to work remotely. Carswell was one of those companies that required most of its employees to work from their homes. It happened almost overnight, said Chief Financial Officer Robert Parsley, but given the situation, we had to provide our employees the chance to work from home and make it as safe and efficient as possible. Conducting the cybersecurity assessment allowed us to transition our IT operations much more quickly, and mobilize faster, added Robert. We were more familiar with what was needed from an IT perspective to continue conducting business remotely in a secure fashion."

Robert Parsley, CFO and Vice President, Carswell Distributing

talent shortage. To add insult to injury, most companies don't even need a full-time CISO. How are modern-day organizations getting their cybersecurity needs met? That's where a virtual CISO like ours comes in.

With our virtual CISO service, clients gain access to a team of certified and highly skilled cybersecurity experts on an "as-needed" basis for a fraction of the cost of a full-time resource. Security is vital for every organization, but not every organization can afford to recruit, hire, and retain a cybersecurity expert. Our team is available "on demand" to serve as trusted advisors who can fulfill any organization's data security needs.

We've made it our objective to work with clients to mature their cybersecurity programs with our experts who work hard and truly care about protecting people's information.

Chris Kolezynski, a Cybersecurity Engineer, is passionate about martial arts and physical fitness. He has studied a variety of fighting styles, including Krav Maga, kickboxing, Kenpo, Muay Thai, boxing, jiu-jitsu, Judo, and more. Chris brought his passion to the ring, competing in national kickboxing tournaments around the country. He is currently training with pro and amateur fighters in preparation for cage fighting for Rising Dragon MMA. Through his vigorous workouts and focus on healthy living, Chris has lost over 200 pounds. He also encourages others to live healthy and strong with motivational posts on social media and his interactions with coworkers.

27

MENU OF PRODUCTS

TCDI provides advanced litigation support software and services for electronic discovery, hosted review and production, and large-scale litigation case file management. The company combines advanced technology and automation with superior client partnerships and has been a technology partner in some of the largest litigation cases in U. S. history.

For corporations, there is always a growing volume of data from year to year that impacts how they manage their legal defenses. With growing volumes, come growing costs and businesses are always seeking to streamline these costs. This process is operated continually from a disparate number of companies, technology, and processes with few providing the streamlined approach where all three entities work together to solve the overarching issues.

With the cooperation of TCDI Project Managers, CVOnyx, and CVLynx, the complete process is managed by bringing together corporate counsel, outside counsel, and technology. CVOnyx processes data with extreme speed and generates exports in load file formats suitable for any major review and production platform. Awarded by *KMWorld* magazine for its innovative design, CVOnyx handles hundreds of different file types and offers flexible, custom processing to meet the requirements of large-scale review and production projects.

TCDI streamlines the entire process and creates a fixed fee arrangement with corporate and outside counsel. Allowing corporations to triage and prioritize litigation, outside counsel to advise solid legal counsel and TCDI to administer the processing and host the data, allowing all three entities to apply their highest and best use to seek resolution of the matter. CVLynx is the data repository where all entities can communicate, create the best-case strategy and execute a solid plan for the duration of the litigation.

CV Fox

The mid-tier corporate and law firm market has been underserviced with technology catered to them to manage electronic discovery. Antiquated platforms create a large barrier to getting their essential data reviews completed and are left to outsource the processing and hosting at a premium cost. This has caused many unnecessary settlements when corporations or firms could otherwise review initially to determine merit.

TCDI believes that ALL cases should be given access to the best technology so they can be properly litigated or defended on merit. The industry has been too self-serving. TCDI sees self-service as providing access to information with simplicity under the cost of traditional discovery.

Data Processing

TCDI's engine, CVOnyx, was built with a modular design that allows us to easily add or replace processing and cull technologies to meet specific or custom processing requirements. Through CVOnyx, TCDI technicians set the processing requirements to run an automated workflow that distributes data to the appropriate processing application in the proper order.

CVOnyx not only offers functional capabilities but also the benefit of speed and scalability by managing the distribution of work over a large farm of "server-soldiers." This allows us to streamline processing based on individual project needs.

All electronic data collected is pre-processed by inventory and filtering modules. Client-defined culling parameters can be applied to the pre-processed data. Only the data that meets a potentially relevant standard moves on to full processing and integration with the CV review and production platform. If required, TCDI can cull data using any metadata or text field available.

At any time if there is a need to vary the parameters used in data filtering, all collected data remains accessible for updating or data sampling. The output of any revised effort once again makes the transition into

CV. If challenged, an automated audit trail is available to document and support the filtering criteria used which provide a defensible record of every document filtered in or out.

This integrated approach to data filtering and reduction gives greater flexibility and lower costs in a manner that is both automated and defensible.

In addition to inventory and filtering, CVOnyx is the engine where all other standard and advanced processing takes place. CVOnyx can filter and eliminate duplicate copies of data. It can also extract metadata and text from more than 390 file types, including Lotus Notes databases and Exchange email, and create load files for all other review platforms currently on the market.

Barbara Dunn, Managing Director in our New York office, was born and raised in The Bronx. After serving as President of the Student Council at St. Barnabas High School and graduating from Iona College, Barbara was an auditor for Arthur Andersen for four years. While at Andersen, she got her CPA license from the state of New York. Barbara left Andersen for an opportunity at Texaco and worked there until the ChevronTexaco merger was completed in 2001. During her tenure at Texaco, she relocated to Houston which was a culture shock for the Bronx native *and* the Texas locals. After 11 years, Barbara returned to her roots in New York. Even though she learned to appreciate the Texas sports teams, Barbara's heart belongs to the Yankees, Giants, and Rangers. When she isn't cheering them on, she enjoys doing Pure Barre and yoga. Additionally, Barbara has had the same crew of girlfriends since grammar school and St. Barnabas. She and her fiancé, Joe, have trips to the Super Bowl, the Kentucky Derby, and Italy on their bucket list.

28

INNOVATIONS

IT Innovation demand happens at warp speed. With our talented industry experts, we keep up with and exceed expectations. Here are a few innovations TCDI developed over the years.

Automated Mailbox "Bin" Processing

Back in the 1990s, documents were scanned from paper to Tiff images. In the early 2000s, we worked with a law firm to develop a process where they could scan and send those documents into coding bins using our CVEdit application. Various law firms emailed files to our coding team. We needed an "Intake" process so the files could be submitted to a coding bin and coded via CVEdit. We automated this process by creating the Intake email addresses per desired bin (i.e., complaints, normal pleading distribution, rush, etc.) and those emails would be submitted to the appropriate drop-boxes where they could be selected, unitized, and coded via CVEdit. This was a game-changer since it allowed law firms to directly submit documents to the appropriate bins for coding. It eliminated time-consuming steps for receiving scanned or paper documents and triaging those documents to the appropriate bins for coding. This has all morphed and expanded over time as new needs and technologies emerged to meet those needs. Over the last few years, we have built-in options that allow users to subscribe to DocketBird, Pacer, etc., and auto-forward or insert an intake email address to the subscription service to automatically submit documents into their Onyx processing environment for loading to their CVLynx site. All documents received are processed where we make the images, natives, and full text available for each of the documents. We complete deduplication upfront before processing and we build in any desired automation/data mapping rules in their Onyx Workflows …. all configurable and all based on specific client needs for the project.

CVEdit – to – Workbench

CVEdit was a robust tool we developed and used with law firms during the early 2000s. However, it was a VB application that required the installation of software on the end-users' workstations, the configuration of user permissions and capabilities, and the structure of the coding templates via INI files that were also placed on the user workstations. This required a skilled person at the clients' end to be able to complete installations and lots of overhead time. Some of that extra time involved managing clients' workstations plus additional troubleshooting. We needed a better way. We developed Workbench, a web-based application, to replace our original CVEdit (VB application). We officially released our first Workbench site in 2016. We needed to move to a web-based application for unitization and coding to provide better support. This created ease of use for our customers plus smoother maintenance and configuration for our team. Workbench not only replaced existing functionality in CVEdit, but it also provided enhanced tools for reporting, unitization, easier coding options, and it utilized CVOnyx for enhanced features for deduplication, full-text extraction, OCR, and native file loading.

Entity Extraction for Coding

In 2017, we released new functionality in Workbench to allow for data entity extraction for names and dates. Previously during document coding, law firm coders experienced lengthy coding times for documents that listed numerous names they wished to identify, validate, and code. One problem was that there were numerous attorney names attached to a pleading as well as witness names, all part of the witness designation list. This made it difficult to ferret out information. Likewise, it was cumbersome to comb through a document to identify dates (document date, date filed, date served) or lists of dates and deadlines from a case management order.

Our solution: We implemented a tool that identifies the dates and names (including names and email addresses) where those values are highlighted (in different colors) in the extracted full-text pane for the document viewer in Workbench. This allows coders to simply click on the names or dates and send them to a coded field. For names, this will also allow the coders to select and "match to" or verify against the names already verified/validated in a Look List. Meaning, if the attorney name and email address have already been historically verified and validated, a coder can select the email address that is highlighted in the text and send it to the Responsible Attorney field. It will display the attorney name in their desired Last Name, First Name format. This is an example use-case, it is all configurable depending on the clients' needs. This was a huge time saver for coders!

Slack Enterprise Review Workspace

In October of 2021, a client requested customization of the Slack messages previously collected, processed, loaded, and produced by a previous eDiscovery vendor. The workspace contained over 50 million single-entry messages. The distinct entries caused an issue with the ability to properly review and determine a string of messages. To effectively review the data, the TCDI teams and client worked together for four months to develop and customize the messages to resolve the issue.

Our CDS team re-processed the original data in CVOnyx and matched the previous coding to the newly processed data. Our SysOps team spent many hours transferring metadata and coding to the newly processed data. Once completed, our Dev department stepped in to develop the module in CVOnyx that would provide the requested information. The module grouped a set of messages in 24-hour increments which we termed Slack Rollups. Our team and the client reviewed demos of how the data was presented. After each demo, the module was updated to deploy the requested changes. Once the final product was determined, we exported the data into the review workspace with customized rollups.

> *Smith Anderson - the largest business and litigation law firm headquartered in the world-renowned Research Triangle region and one of the largest in North Carolina*
>
> *"The MSMR program has provided us the opportunity to work with dedicated, hardworking attorneys. I personally have worked on numerous complex reviews with the program – the flexible and creative approach to those reviews have been invaluable in our service to our clients, and the results have been consistently outstanding. I've personally enjoyed working with Jenn over the last several years. I found an opportunity to develop that working relationship across multiple matters and with multiple clients to be invaluable in establishing a collaborative approach between us. It's a truly unique and major benefit of the program in my view. We, at Smith Anderson, have enjoyed watching the MSMR program grow from its infancy to where it is today."*
>
> *Andrew Atkins, Smith Anderson*

Public Websites

In 1998, Altria was required by the Tobacco Master Settlement Agreement to create a public website for all tobacco documents that anyone could access. TCDI began hosting it in 2008 and created Philip Morris Documents (PMDOCS) and Philip Morris Litigation Documents (PMLITDOCS). PMDOCS was for anyone and hosted public documents, Privilege Log Index, and Confidential Document Index. Registered users used PMLITDOCS to access public and confidential documents.

Initially, another company was hosting this data, but with the data transfers and work that it generated, it made more sense for TCDI to host the data in 2010. The look and feel of the website were controlled by the settlement, so we wrote a custom version of CVLynx to fit the requirements.

Over time, some of the requirements have changed and TCDI has been able to easily accommodate those variations. In 2022, the settlement expired. This required additional custom modifications as PMDOCS is now 'frozen' with no adjustments to be made while PMLITDOCS remains active and changes are updated.

Hosting the public site (in addition to generating additional revenue for TCDI) was a great model for companies with mass litigation to provide a place from which opposing parties can search and pull data from the sites. For every matter brought against them, they did not have to do a full discovery request. Their entire past set of relevant documents was already available.

Data Loader

Throughout the life of a project, there is a need to update certain metadata fields in a bulk capacity. These options existed but for one field at a time or for multiple fields to be updated with the same values. As different matters became more complex, there was a need to update multiple fields across multiple records with different values.

In 2012, Data Loader was developed to serve that function and create exceptional value for our clients. It allows users the capability to run updates that previously were sent to the database administrators requiring time lost to create tickets and for those tickets to wait in the queue to be completed. Our Project Managers are now more responsive to user requests.

In addition, Data Loader validates that the fields being updated are in the correct format. This is critical when updating data fields. The tool has evolved over time and now includes adding records to the database instead of only updating, appending values, and adding static values to existing fields. As a result, data is protected against incorrect updates by guarding extracted metadata fields while providing access only to user-updated fields.

Review

Around 2014, TCDI noticed that document review styles were changing to become more efficient. 'Traditional' review was to review full families from the start of the review at each step. This changed to reviewing only the documents that hit the agreed-upon search terms.

Our development team created an automated way to accommodate both styles of review using CVLynx custom review workflows. A family review can occur if that is what our client intends to do. When our client wants to review direct hits only, there needs to be a way to bring in the family members of documents that were coded as Responsive.

This step in the workflow prevented manual manipulation of the documents in review and created efficiency for the review teams and their admins. Responsive documents would have family members added automatically so that a full review of the family could be completed. Non-responsive documents would exit review.

Poduction Wizard

Production is our final delivery. It's what gets presented to our clients while maintaining accuracy and security. In the early years, the tool in CV required a lot of updates and a manual list of modifications. The Development Team collaborated with a Senior Project Manager to rewrite the steps and run them automatically. Our production tools now eliminate the risk of errors and are more streamlined.

W̶hen we received a large list of product-related terms, we processed them and combed through the data using specific terms. During the analytical step, the database volunteers search term suggestions in addition to those terms. This refining process captured and tagged the documents that mattered the most. Any documents that weren't tagged could be deleted.

Data Mining Analyst **Angelus Jimenez** had been a contractor for our MSMR Program for four years and came on full-time to TCDI in September 2022. Being a military spouse for the past 20+ years and a mom of three has not only been challenging but career-halting for her. Before coming to TCDI, Angelus worked the night shift in a Neonatal Intensive Care Unit while her husband was deployed. For 15 years, she juggled a full-time job, kid schedules, and made sure her Autistic son attended ABA, speech, occupational, physical, water, and feeding therapy several times a week. When she resigned from the NICU, Angelus focused on her kids and supported her husband's military career. After a while, she started volunteering through her kids' school PTO and then at Murphy Canyon Bread Ministry in San Diego, CA. She found her calling to volunteer and help military families in need by providing breads, pastries, non-perishables, produce, refrigerated goods, dairy, and meats seven days a week. Angelus soon became Assistant Director for the next three years. She expanded the program by adding almost 60 military spouses or active duty volunteers, picked up 10 more donating stores, and, most importantly, got at least one bread site in several military housing communities. More recently, Angelus and her husband started the For the Ones who Serve non-profit to provide more programs for active duty, retirees, and veterans.

29

SECURING DATA BEYOND THE OFFICE

Data Security, Privacy, and Confidentiality are the priority concerns of our clients. While TCDI is known for its compliance with strict information security policies including Systems and Organization Controls 2 (SOC2), Privacy Shield, General Data Protection Regulation (GDPR), and HIPAA, building the foundation for a secure remote workforce was not as simple as giving reviewers credentials to the review site and hoping for the best.

Fast forward to 2020 when businesses across the country were forced to transition to a work-from-home model. Many successful companies scrambled to set up platforms for remote review in hopes their clients would be comfortable with a quick-fix approach to secure their data in a remote environment.

TCDI already had a platform in place to allow for safe and secure remote document review. Policies were vetted and audited by current clients. TCDI had supporting documentation and audit material for new clients who also needed to continue with business as usual.

The remote review program was created in 2017 using a Virtual Review Center (VRC) concept. VRC ensures that review will be completed in a highly secured environment, mirroring an on-site review center. VRC accomplishes this by operating as a multi-layer approach to security. VRC also simultaneously grants access to a geographically diverse set of reviewers with full data protection.

A few security highlights of the remote review program include:

1. Our network and data center infrastructure are designed with industry-standard best practice controls, ensuring all access is documented, approved, monitored, and managed to ensure only authorized users and services access the network resources. In addition to perimeter controls, we also utilize internal firewalls to protect data transfer and access between client environments.

2. All reviewers access documents through secure Virtual Machine (VM) terminal access. This allows us to set up secure, locked-down access controlled by our Information Security team that only allows users with necessary access to complete their job.

3. Every project is also set up to mimic a Virtual Classroom Environment and every reviewer on the project is assigned to that classroom. Project Managers can view each team member's VM session LIVE to monitor activity and assist the team. Additionally, the Project Manager can communicate with individual reviewers or the entire classroom (Review Team). On top of that, the PM can end any session at his or her discretion, as well as perform virtual trainings through this classroom by displaying screens to all reviewers simultaneously. Not only does this add a necessary security level, but it also broadens our ability to communicate and collaborate with our review team.

4. VRC's security design ensures authentication and protection protocols meet or exceed those of our on-site review centers. As such, the virtual environment fully integrates into the information security architecture used in the TCDI data center. Access to our data center resources is controlled by various two-factor authentication methods, with access controlled by RSA's SecurID or RBA protocols. Internal server resources are further controlled by two-factor authentication systems.

5. TCDI maintains a centralized Security Information and Event Management (SIEM) platform for threat detection monitoring. We additionally manage from multiple sources, including, but not limited to: firewalls; web gateway; email gateway; Intrusion Prevention System (IPS); application and network services; and Advanced Malware Prevention (AMP).

TCDI has always strived to stay abreast of workforce changes and influences.

30

AUTHOR'S PERSONAL PERSPECTIVE

In July 2022, I received an email from a former coworker whom I hadn't had any contact with for a good 18 months. I always thought of her as one of the smartest people I had ever met. After all, she had once worked for NASA. In her email, she wrote, "My manager asked me if I knew of any good business analysts and I immediately thought of you." I was immediately flattered. She wrote a few sentences about her company while giving me highlights on the great people, the work, and the environment. TCDI was now on my radar.

Her timing was perfect because I was quietly looking for a new job. A couple of weeks later, I interviewed with her manager. My extroverted personality could be intimidating so I wasn't sure how the interview went. A few days later, he verbally offered me the job. All this time, I had been concentrating on my BA skills and experience and what I could bring to TCDI, not knowing the depth and breadth of the company.

I started at TCDI on September 16, 2022, working remotely from Pennsylvania.

When Bill heard that I had written 13 books, he wanted my first project to be to pen TCDI's success story. Most of my books are fiction novels so a non-fiction corporate biography was not in my wheelhouse. Needless to say, I felt daunted. Not only was I a newbie at TCDI, but Bill wanted me to compile a book that was not my usual genre. But I didn't let these challenges deter me. I then interviewed Bill and several of the senior managers at TCDI to learn the unique story behind our company. Everyone was willing to talk to me and the corporate culture impressed me. Everybody seemed genuinely happy and proud to work for TCDI.

A few weeks into my new job, Bill contacted me and asked if I read any books by *New York Times* best-selling author Elin Hilderbrand. With an embarrassed grimace, I admitted I had not. Bill invited me to Greensboro to meet Elin because she was friends with Bill's wife, Susan. Elin had plans to travel to North Carolina from Nantucket for a breast cancer fundraiser with earlier.org that Susan and I were attending.

Remembering previous business travel, I figured I would have to do the initial legwork of finding a flight and a hotel. Easy enough. With my love for travel, I had mastered online booking services. Then Bill threw me a curveball. He asked me what my closest airport was so that he could send the TCDI jet to pick me up. Much to my chagrin, the first words out of my mouth were, "Holy crap!" I had never experienced working for a company where making sure a new employee's transportation was furnished like this. (Full disclosure, I had worked for state government for 18 years and was lucky to get my mileage reimbursed.) An eight-passenger mini-jet met me at my nearest airport and I didn't have to worry about the hassles of checking bags, waiting through security, or rushing to make the plane's takeoff time. This new concept blew me away and made me, a brand-new employee, feel welcomed and appreciated. I couldn't think of another small company or CEO that would do this.

On that same day, I received some heartbreaking news from my veterinarian. My beloved 13-year-old American Staffordshire Terrier, Charlotte, had an inoperable liver tumor. She was in good health--for the time being. The vet told me she could have as little as six weeks left. That night at dinner with Bill, I mentioned my pooch's dire diagnosis. With tremendous empathy, he told me, "I'll have the pilot on standby to run you home tomorrow to be with your puppy… priorities." His kindness and sincerity blew me away. No other boss had ever treated me like this. (When my grandmother died, my boss at the time complained that I took time off for the funeral.) I didn't need to rush home to Charlotte, but I appreciated Bill's generosity. He is one of a kind.

Welcome to the TCDI family.

Charlotte passed away at home on March 4, 2023, five months after her liver tumor was discovered.

Contributors

Bill Johnson

Janet Hamilton

Geoff McPherson

Dave York

April Marty

Sandy Middlebrook

Joe Anguilano

Elizabeth Wagoner

Dave Martyn

Jennifer Andres

Ned Adams

Laura Hale

Nancy Johnson

Andy Cosgrove

Lisa Mays

Acknowledgements

TCDI wouldn't be where it is today without our clients, past and present.

We thank you from the bottom of our hearts helping us succeed.

ABOUT THE AUTHOR

Mary Walsh is an award-winning multi-genre author of 15 books: non-fiction, crime, urban fantasy, historical fiction, cozy ghost story, and romance. Her latest novel, *The Curse of Jean Lafitte*, a paranormal adventure featuring Jean Lafitte the Pirate in modern-day New Orleans, dropped on August 31, 2023. She lives in Mechanicsburg, PA, with her husband Dave and their four kids.

Order autographed books at:

marywalshwrites.com

Follow her on Goodreads and Amazon:

www. goodreads.com/goodreadscommarywalshwrites

www. amazon.com/author/marywalsh2